Book 42, the connections between math function, prose, poetry, and the storytelling system

The art of creating concise messages using compactification fundamentals

This book celebrates the connections between math function, prose, poetry, and the art of creating concise messages using compactification fundamentals.

It covers scalable stories, starting from no words or symbols such as non-existence, nothingness, absence of all quality or quantity, void of non-being to 1 to 299 words considered a communication sweet spot scalable range, then scale up to 1,500, 5,000, 20,000, 40,000, and beyond.

Word Symbolism:

Minus = Debt, hidden, burden, drag

0 = non-existence, nothingness, absence of all quality or quantity, void of non-being

1 = Primordial unity, the beginning, the creator, sum of all possibilities, essence, the center, the indivisible, the germinal, isolation, an upsurging and uprising

2 = Duality, alternation, diversity, conflict

3 = Multiplicty, creative power, forward movement, overcoming duality, all is appropriated, power of 3

4 = Solid figure, special order, the divine quaternity, static, compared to dynamic

5 = The human microcosm, the pentagon, wholeness, the *hieros gamos*

6 = Equilibrium, the perfect number

7 = The number of the universe, a macrocosm, completeness, perfection, security, rest, plenty, reintegration, synthesis, the great mother

8 = Spiritually, initiate, next stage, paradise regained, regeneration, perfect rhythm

9 = Composed of the all-powerful 3 x 3, triple triad, completion, fulfillment

10 = The number of the cosmos, all possible

11= Danger

12 = Duodecad, both spiritual and temporal order

50 = Jubilee, return to the beginning

And this book explores how to craft quintessential phrases, elevator pitches, loglines, taglines, microfiction, micrononfiction and more.

And this book guides you through the process of creating messages that are clear, effective, practical, and vivid based on Vitruvian principles to help solve daily problems, explain the universe, and even lead to an eternal paradise …

… all with a money-back guarantee?

***

And ask any writer to cut a sentence, paragraph, chapter, or book by 10 percent, 20 percent, or more.

The thought of hacking and hacking and hacking at that book or whatever like a barbarian or typical savage could drive that writer to spark and bark about in public.

Yes, or no?

As our mission, at least mine, Bryan Fletcher, bfletch157@yahoo.com is to restore order and clearly describe that process, and guide the viewer to a much better, sustainable place, then improve that, then open that place to fresh start in a good homestead region with fresh air, water, food, shelter, and sublime view that shows a series of mountains, valleys, streams, river, coast, sailboat, and options to explore with a series of practical ways, and if need be, sail on, sail away on a truly great adventure, and return with a boon, and repeat that process.

True or false?

Original content

Book series written by Bryan Fletcher, born in Princeton, NJ, USA, now living in the wonderful, inexpensive city of Bluefield, WV, USA, bfletch157@yahoo.com. and website to support each book series: https://website4384700.nicepage.io/About-the-author.html, and https://www.amazon.com/stores/Bryan-Fletcher/author/B07N6PM6Y9?ref=ap_rdr&store_ref=ap_r dr&isDramIntegrated=true&shoppingPortalEnabled=true

And Bryan is also known by other names, yet some we cannot say in public:

especially to a very impressionable teenager reader, aka a typical juvenile or adult version, or family, tribe, or other social groupoid equivalent stuck in an adolescent phase or worse.

However, it is fairly accurate to say, Bryan has been called David, and David Kerr, and Steve Kerr, Steve Raichlen, David Raichlen, as well as, a subpar polyliterary polyglot microspeak descriptive specialist and pragmatica scripturient, bibliophile, Mersenne prime function, logophile, owitic, bibliolater, abiblio, bookarazzi, book-bosomed, bibliosmia, librocubicularist, scrollmate, bibliosmia, aeonic procatalepsis scholastican storytelling nerd literaryphiloic, subpar polyliterary, reasonable textualist, and reasonable Archaebacteria Eubacteria Protista Fungi Plantae Animalia storytelling specialist, a thoughtful thinkculum that seeks $f(x) = \int$ Sublime Change(x) dx.

And Bryan is known as an illegitimate offspring of Marcus Vipsanius Agrippa, Calvin Euler, Sir Isaac Newton, Andrzej Tadeusz Bonawentura Kościuszko Agrarianism, James Augustine Aloysius Joyce, Alan Turing and more quintessential troublemakers, aka rascals that help create original content adventures.

Book editor: none yet, as he may balk.

Because this is a raw notebook version filled with a large number of original contents as well as classic ideas.

CHAPTER 1: This book celebrates the connection between math function, prose, poetry, and the art of creating concise messages using compactification fundamentals.

And it covers scalable stories

CHAPTER 2: Math, prose, poetry as well as other ways and means may seem like distinct realms?

CHAPTER 3: A classic situation, look for love, find, ends badly, escapes, then recover, and grow better

CHAPTER 4: Series of Sonnets

CHAPTER 5: Literary criticism in the field of math, then show why, and offer a summary of excellent ideas to judge a math story, aka an expression

CHAPTER 6: Math and storytelling may seem like two entirely different realms?

CHAPTER 7: Phonemes, structure and patterns

CHAPTER 8: The relationship between math and intonation

CHAPTER 9: Stress in linguistics refers to the emphasis placed on certain syllables or words often describes using mathematical concepts like patterns and probability

CHAPTER 10: The relationship between math and storytelling morphology sets in the structured, rule-based systems they both utilize

CHAPTER 11: Imagine mathematics and prose syntax as two master chefs in a bustling kitchen

CHAPTER 12: Mathematics, semantics, and storytelling may seem like distinct fields

CHAPTER 13: Mathematics, pragmatics, and storytelling might seem like separate realms

CHAPTER 14: The Relationship Between Math and Orthography in Storytelling

CHAPTER 15: Imagine mathematics and the Prosody storytelling system as two master architects working together to design a magnificent building

CHAPTER 16: Start with a complete story, then transform it into proofs, mathematical functions, epic poems, folk guitar songs, and popular TV episodes

CHAPTER 17: Microfiction stories with less than 299 words and those compactification attempts

CHAPTER 18: Symbols used in both math and

language, then sort symbols by purpose or function

CHAPTER 19: Microfiction story (299 words)

CHAPTER 20: Highlights the relationship of math to finding a new practical series of idea for a fresh sustainable start as well as those ways and means do and find wealth and easily grow and sustainable system then build a truly great nation

CHAPTER 21: Highlights the relationship of math to finding a new practical series of idea for a fresh sustainable start for a new epic forest adventure

CHAPTER 22: Find a great treasure map and prep then go on that fresh adventure

CHAPTER 23: The relationship of math to finding a sustainable happiness

CHAPTER 24: Highlights the relationship of math to finding an excellent husband

CHAPTER 25: Highlights the relationship of math to finding a new practical to find the best community to live in

CHAPTER 26: Highlights the relationship of math how to win a major cooking contest

CHAPTER 27: Highlights the relationship of math and create the perfect pillow and bed

CHAPTER 28: Highlights the relationship of math to music theory

CHAPTER 29: Last year, I had an experience that completely changed my outlook on life.

I wanted to finish a marathon to prove to my kids that anyone can do it one week before the big run, I slipped and injured my wrist.

Completely shocked, I stood up, looking at my …

CHAPTER 30: Would you like a truly great adventure?

Yes, or no?

OK. OK. OK.

Here are a few easy folk guitar songs to play during your next truly great adventure, yes, yes, yes, and along with their chords that tell an excellent math function story proof of concept, as if a universal function?

True or false?

And feel free to vote.

CHAPTER 1

This book celebrates the connection between math function, prose, poetry, and the art of creating concise messages using compactification fundamentals.

It covers scalable stories, starting from no words or symbols such as non-existence, nothingness, absence of all quality or quantity, void of non-being to 1 to 299 words, considered a communication sweet spot range, then scale up to 1,500, 5,000, 20,000, 40,000, and beyond.

***

And ask any writer to cut a sentence, paragraph, chapter, or book by 10 percent, 20 percent, or more.

The complex thought of hacking and hacking and hacking at a likely delusional masterpiece could drive that writer to bark about: especially in public.

Yes, or no?

Well, at least for me, it seems that way.

"Cut what?

"Cut where?

"Cut out an event, or character, or style, or what?

"This is a vague request or whim.

"And that seems as if yet another fool's errand.

"And what?

"Do you want me to hack a 40,000-word adventure novel into less than 1,500 words?

"Are you insane?

"It's impossible.

"Well, possible, yet I'm not interested in yet another fool's errand."

***

And this book could offer several additional benefits, such as:

1. **Inspiration for Creativity**: By blending math, microfiction, and poetry, it can spark new ideas and creative approaches in both writing and problem-solving.
2. **Enhanced Communication Skills**: Learning to craft concise messages like elevator pitches and taglines can improve your ability to communicate complex ideas clearly and effectively.
3. **Interdisciplinary Learning**: It can provide insights into how different fields intersect, fostering a deeper appreciation for both the arts and sciences.
4. **Mental Agility**: Engaging with compact and precise forms of expression can sharpen your thinking and

improve your ability to distill information, a quintessential communication skill.

5. **Educational Tool**: It can serve as a unique resource for teaching concepts in both mathematics and literature, making learning more engaging and enjoyable.

6. **Personal Growth**: Exploring the relationship between different forms of expression can lead to personal insights and growth, enhancing your overall intellectual and emotional well-being.

***

And this book could indeed show how mathematical formulas can inspire and structure prose, phrases, sentences, chapters, and entire stories. Here's a detailed explanation along with some vivid examples:

How Math Formulas Can Inspire Writing

1. **Prose Word**:
   - **Example**: The Fibonacci sequence (0, 1, 1, 2, 3, 5, 8, …) can inspire the creation of words by using the number of letters in each word to match the sequence.
   - **Illustration**: "A" (1 letter), "is" (2 letters), "an" (2 letters), "art" (3 letters), "of" (2 letters), "words" (5 letters).

2. **Phrase**:
   - **Example**: Using the Golden Ratio (approximately 1.618) to balance the length of words in a phrase.
   - **Illustration**: "Harmony in nature" (5 letters, 2 letters, 6 letters - approximating the ratio).
3. **Sentence**:
   - **Example**: Applying the concept of symmetry from algebra to create a balanced sentence.
   - **Illustration**: "The sun rises in the east and sets in the west."
4. **Chapter**:
   - **Example**: Structuring a chapter using the concept of fractals, where each part mirrors the whole.
   - **Illustration**: A chapter that starts with a small story, expands into a larger narrative, and then concludes by reflecting back on the initial story.
5. **Story**:
   - **Example**: Using the structure of a mathematical proof to outline a story, with a

hypothesis, supporting arguments, and a conclusion.
- **Illustration**: A mystery novel where the protagonist formulates a hypothesis about the crime, gathers evidence, and finally proves the hypothesis in the climax.

Poems Inspired by Math

1. **Haiku Using Pi**:
    - **Example**: A haiku where the syllable count of each line follows the digits of Pi (3.14).
    - **Illustration**:
    - Infinite circle,
    - Three point one four, endless sky,
    - Nature's mystery.

2. **Fibonacci Poem**:
    - **Example**: A poem where the number of syllables in each line follows the Fibonacci sequence.
    - **Illustration**:
    - One,
    - Small,
    - Step,
    - Forward,
    - Giant leap,

- For mankind, we soar,
- Beyond the stars, we explore.

Epic Folk Songs for Guitar

1. **Song Inspired by Euler's Identity**:
    - **Example**: A song that celebrates the beauty of Euler's identity: (e^{i\pi} + 1 = 0).
    - **Illustration**:
    - (Verse)
    - In the realm of numbers, a beauty unfolds,
    - Where \(e \) and \(\pi \) and \(i \) are bold,
    - They dance together in a cosmic ballet,
    - And bring us Euler's magic display.
    -
    - (Chorus)
    - Oh, Euler's identity, so pure and true,
    - A symphony of math, in skies so blue,
    - With \(e \) and \(\pi \) and \(i \) in harmony,
    - They sing the song of infinity.

2. **Song Celebrating the Golden Ratio**:
    - **Example**: A folk song that tells the story of the Golden Ratio and its presence in nature and art.
    - **Illustration**:

- (Verse)
- In the petals of a flower, in the spiral of a shell,
- The Golden Ratio whispers, a story it does tell,
- From the pyramids of Egypt to the stars up in the sky,
- This ancient number guides us, as the ages pass by.
-
- (Chorus)
- Oh, Golden Ratio, in nature's grand design,
- You weave a thread of beauty, through space and time,
- In the art of the masters, in the waves of the sea,
- The Golden Ratio's magic, forever will be.

***

And these examples show how mathematical concepts can be creatively applied to various forms of writing and music, making the book a rich resource for both learning and inspiration.

***

Or show another way:

Math Functions

1. Relationship of Math to Storytelling:
   o (f(x) = \text{compactification}(x))
   o (g(x) = \text{quintessential phrase}(x))
   o (h(x) = \text{elevator pitch}(x))
   o (i(x) = \text{logline}(x))
   o (j(x) = \text{tagline}(x))

2. Scalability of Messages:
   o (S(x) = \text{scale}(x, \text{microfiction}, \text{epic novel}))

3. Editing and Refinement:
   o (E(x) = x - \text{unnecessary content})

4. Mission Statement:
   o (M(x) = \text{guide}(x, \text{better place}))

***

And is there a relationship between math, words, sentences, paragraphs, and the storytelling system?

***

Absolutely, math is often considered a language. It has its own set of symbols, syntax, and rules, much like any spoken or written language.

Just as we use words and grammar to communicate ideas in English or Japanese, we use numbers and mathematical operations to express relationships and solve problems in math.

It's a universal language that transcends cultural and linguistic barriers, allowing people from different backgrounds to understand and communicate complex concepts.

So, in a way, math is a language that speaks to the logic and structure of the universe.

And music is a language type or form.

And regarding language types, there is the prose system, aka the Latin expression *prosa oratio*, literally, straightforward, direct, and natural speech or chat of a typical person, as compared to another language type, poetry.

Poetry is a literary art form that employs aesthetic and rhythmic qualities of language to convey deeper meanings beyond the literal. It often uses metaphor and other devices to describe complex ideas or mysteries.

And it is characterized by its use of various poetic devices and structures to evoke emotions, create imagery, and express complex ideas.

And here are some key elements and techniques associated with poetry:

Key Elements of Poetry

1. Aesthetic and Rhythmic Qualities:
2. Poetry often uses rhythm, meter, and sound patterns to create a musical quality. This can include the use of rhyme, alliteration, assonance, and consonance.
3. **Poetic Devices:**
    - **Assonance:** The repetition of vowel sounds within words to create internal rhyming.
    - **Alliteration:** The repetition of consonant sounds at the beginning of words.
    - **Euphony:** The use of pleasant, harmonious sounds.
    - **Cacophony:** The use of harsh, discordant sounds.
    - **Onomatopoeia:** Words that imitate natural sounds.
    - **Metaphor:** A figure of speech that compares two unlike things without using "like" or "as."
    - **Simile:** A figure of speech that compares two unlike things using "like" or "as."
    - **Symbolism:** The use of symbols to represent ideas or qualities.
4. **Verse and Structure:**

- Most poems are formatted in verse, which consists of a series of lines arranged in a specific rhythmic or metrical pattern. This structure distinguishes poetry from prose.
- **Stanza:** A grouped set of lines within a poem, often separated by a space. Stanzas function similarly to paragraphs in prose.
- **Meter:** The rhythmic structure of a poem, determined by the number and arrangement of stressed and unstressed syllables.

5. **Forms and Conventions:**

    - Poetry uses various forms and conventions to suggest different interpretations of words and evoke emotional responses. These forms can include sonnets, haikus, odes, and free verse.

6. **Ambiguity and Multiple Interpretations:**

    - Poetic language often employs ambiguity, allowing for multiple interpretations. This can be achieved through the use of figurative language, irony, and other stylistic elements.

7. **Historical and Cultural Context:**

- Poetry has a long and varied history, evolving differently across cultures and time periods. It can reflect the values, beliefs, and experiences of a particular society.

***

And a writer of proses should carefully borrow these when situationally appropriate.

***

And just as importantly, math and prose language are more intertwined than you might think.

Because both are systems of symbols and rules used to convey meaning.

In math, the system uses numbers and operations to express relationships and solve problems.

In language, we use words and grammar to communicate ideas and tell stories.

Words and sentences in language can be seen as analogous to numbers and equations in math.

Just as numbers can be combined in various ways to form equations, words can be combined to form sentences.

And sentences, in turn, can be combined to form paragraphs, much like how equations can be combined to form more complex mathematical expressions.

And storytelling, like math, follows certain structures and patterns.

For example, a story typically has a beginning, middle, and end, similar to how a mathematical proof has a hypothesis, a series of logical steps, and a conclusion.

And both storytelling and math require logical thinking and creativity to construct meaningful and coherent narratives or solutions, yet there are exceptions.

And in essence, math and language are both tools for understanding the world, each with its own unique set of symbols and rules.

And they complement each other and can be used together to enhance our understanding and communication of complex ideas.

***

Or say in greater detail regarding the the fascinating world of math as a language.

At its core, math is a system of symbols and rules us ed to convey meaning, much like any spoken or written language.

It has its own alphabet of numbers and symbols, grammar (operations and equations), and syntax (the order in which perform operations).

Just as we use words to form sentences and paragraphs, we use numbers and symbols to create equations and expressions.

One of the most intriguing aspects of math as a language is its universality.

While spoken languages can vary widely across cultures and regions, math remains consistent.

A mathematical equation holds the same meaning whether you're in Tokyo, New York, or Timbuktu.

Because this universality allows people from different backgrounds to communicate complex ideas and solve problems together.

And also, math has its own form of storytelling.

In a mathematical proof, we start with a hypothesis, follow a series of logical steps, and arrive at a conclusion.

This process is similar to the structure of a story, which typically has a beginning, middle, and end.

Both require logical thinking and creativity to construct meaningful and coherent narratives or solutions.

Moreover, math can be incredibly poetic. Consider the elegance of Euler's identity: $e^{i\pi}+1=0$. This simple equation connects five of the most important numbers in mathematics (e, i, π, 1, and 0) in a beautifully concise way.

It's like a haiku of the mathematical world, capturing profound meaning in just a few symbols.

In essence, Fi speaks to the logic and structure of the universe. It allows us to describe patterns, make predictions, and understand the world around us in a precise and systematic way. And just like any language, it has its own beauty and elegance that can be appreciated by those who take the time to learn it.

So, whether you're solving a complex equation or crafting a compelling story, remember that both math and language are tools for making sense of the world.

And who knows, maybe one day you'll find yourself writing a mathematical sonnet or composing a great numerical symphony.

CHAPTER 2

Math, prose, poetry as well as other ways and means may seem like distinct realms.

***

Yet they share a deep and intricate relationship.

Because all are systems of symbols and rules used to convey meaning and structure.

So, how does math relate to creating a sentence with some detailed examples?

Structure and Syntax

1. Syntax and Grammar: Just as math has rules and syntax for forming equations, language has grammar rules for constructing sentences. These rules ensure that sentences are coherent and convey the intended meaning. For example, in English, a basic sentence structure follows the Subject-Verb-Object (SVO) pattern.

Example:

- **Math:** $a+b=c$
- **Sentence:** "The cat (Subject) chased (Verb) the mouse (Object)."

Patterns and Sequences

2. Patterns in Language: Mathematical patterns can be found in the rhythm and flow of sentences. For instance, the use of parallelism in writing creates a pattern that enhances readability and aesthetic appeal.

Example:
- **Math:** Arithmetic sequence $a, a+d, a+2d, a+3d, \ldots$
- **Sentence:** "She likes reading, writing, and drawing."

Balance and Symmetry

3. Symmetry in Sentences: Symmetry in math refers to balanced proportions, which can be applied to sentences to create harmony and coherence. Balanced sentences often use parallel structure to achieve this effect.

Example:
- **Math:** Symmetrical equation $x=y$
- **Sentence:** "To err is human; to forgive, divine."

Ratios and Proportions

4. Proportions in Language: Ratios and proportions in math can be applied to the length and complexity of sentences. Varying sentence lengths can create a pleasing rhythm and maintain the reader's interest.

Example:
- **Math:** Ratio ab

- **Sentence:** "He ran. She walked. They strolled together."

Detailed Examples

Example 1: Fibonacci Sequence in Sentences the Fibonacci sequence (0, 1, 1, 2, 3, 5, 8, 13, ...) is a series of numbers where each number is the sum of the two preceding ones. This sequence can be applied to the length of sentences in a paragraph to create a natural and engaging progression.

Example 2: Golden Ratio in Sentence Structure, the golden ratio (approximately 1.618) is a mathematical ratio often found in nature and art. In writing, this ratio can be used to determine the ideal length of sentences and paragraphs, creating aesthetically pleasing and harmonious text.

Example 3: Fractals in Narrative Complexity, Fractals are complex patterns that are self-similar across different scales. In writing, fractal structures can be used to create intricate and layered sentences, where sub-clauses mirror the main clause, adding depth and complexity.

Conclusion

Math and language are deeply interconnected, with mathematical principles providing structure, balance, and depth to sentences.

By understanding and applying these principles, writers can create more engaging and meaningful sentences.

Whether it's through syntax, patterns, symmetry, or proportions, math offers valuable tools for crafting compelling sentences, yes, or no?

***

Or, feel free to offer other ideas, or close this storytelling math proof of concept: *Book Number 42*, then shift to another channel.

Such as, sail on.

***

Yes.

Sail away.

***

CHAPTER 3

Wow, to find love and then, if in a very bad relationship, escape, celebrate newfound freedom, remake yourself into a better person, and show the results of that transformation.

***

Math Equivalent of Words, Phrases, Sentences, Paragraphs, Chapters, and the Entire Story

1. **Words:**
    - In math, words can be represented as variables or constants. Each word has a specific meaning or value, similar to how variables hold values in mathematical equations.
    - **Example:** Let w represent the word "love."
2. **Phrases:**
    - Phrases can be seen as expressions, which are combinations of variables and constants connected by operators.
    - **Example:** The phrase "true love" can be represented as w_1+w_2, where w_1 is "true" and w_2 is "love."
3. **Sentences:**

- Sentences are analogous to equations or inequalities, where expressions are combined to form a complete thought.
- **Example:** The sentence "True love conquers all" can be represented as $w1+w2=w3$, where $w1$ is "true," $w2$ is "love," and $w3$ is "conquers all."

4. **Paragraphs:**
   - Paragraphs can be seen as systems of equations, where multiple sentences (equations) are combined to convey a more complex idea.
   - **Example:** A paragraph describing a romantic relationship might include equations representing love, trust, and communication.

5. **Chapters:**
   - Chapters are analogous to sets of systems of equations, where multiple paragraphs (systems) are combined to form a coherent section of a story.
   - **Example:** A chapter in a novel might explore different aspects of a relationship, each represented by a system of equations.

6. **Entire Story:**
    - The entire story can be seen as a comprehensive mathematical model, where all chapters (sets of systems) are combined to form a complete narrative.
    - **Example:** A novel about love and relationships can be represented by a complex model that includes variables, expressions, equations, and systems.

Math Stories

Story 1: Finding Love

In the bustling city of Numeria, a mathematician named Leo was determined to find love. He believed that math could help him make the best choice. Leo created a mathematical model to evaluate potential partners, assigning values to qualities such as kindness, intelligence, and shared interests.

Math Representation:
- Variables: K (kindness), I (intelligence), S (shared interests)
- Equation: $L = K + I + SC$, where L represents love and C represents compatibility.

Leo tested his model by going on dates and evaluating each potential partner using his equation. He tracked his satisfaction levels and found that the equation accurately predicted his happiness with each partner. Eventually, Leo met Emma, who scored the highest on his model. They fell in love and lived happily ever after.

Story 2: Escaping a Bad Relationship

After years of being in a toxic relationship, Maya decided it was time to leave. She used math to analyze the factors contributing to her unhappiness and create a plan to escape.

Math Representation:

- Variables: T (toxicity), S (stress), H (happiness)
- Equation: $U = H - (T+S)R$, where U represents unhappiness and R represents resilience.

Maya calculated her unhappiness score and realized that the relationship was draining her. She created a step-by-step plan to leave, focusing on building her resilience and reducing stress. With determination and support from friends, Maya successfully escaped the toxic relationship.

Story 3: Celebrating Newfound Freedom

After leaving her toxic relationship, Maya felt a sense of liberation. She decided to celebrate her newfound freedom by exploring new hobbies and interests.

Math Representation:

- Variables: F (freedom), E (exploration), J (joy)
- Equation: $H = F + E + J$, where H represents happiness.

- Maya tried different activities, such as painting, hiking, and dancing. Each new experience brought her joy and helped her rediscover herself. She felt happier and more fulfilled.

CHAPTER 4

Series of Sonnets

Sonnet 1: The Math of Words

In words we find a structure pure and clear, Each letter placed with purpose, thought, and care. Like numbers in a sequence, they appear, To form a sentence, meaning to declare.

The syntax rules, like math, do guide our way, Ensuring clarity in what we write. With patterns, rhythms, guiding what we say, Our prose becomes a beacon in the night.

In paragraphs, like chapters, stories grow, Each sentence building on the one before. With math's precision, narratives do flow, Creating tales that readers will adore.

So let us blend the art of words and math, To craft a story's ever-winding path.

Sonnet 2: The Math of Love

In love, we find a formula so sweet, A balance of emotions, hearts entwined. With math, we measure every heartbeat's beat, And find the perfect partner, love defined.

Equations guide us through the maze of hearts, Predicting joy and sorrow, highs and lows. With every step, a new equation starts, And love's true path, through math, it clearly shows.

But when love falters, math can guide us still, To find the strength to leave, to start anew. With numbers, we can measure every thrill, And find the courage to bid love adieu.

So let us trust in math to guide our way, Through love's bright dawn and through its darkest day.

Sonnet 3: The Math of Transformation

In transformation, math does play a part, A guiding hand to shape our path anew. With every change, a new equation's start, And through its logic, we find what is true.

From sorrow's depths, we rise with math's clear light, To find the strength to heal, to grow, to thrive. With every step, a new and hopeful sight, And through its wisdom, we become alive.

In every change, a pattern we can see, A sequence leading to a brighter day. With math, we find the path to truly be, And in its guidance, we will find our way.

So let us trust in math to lead us on, To transformation's light, from dusk to dawn.

Series of Epic Poems

Epic Poem 1: The Journey of Words

In the land of letters, where stories are born, A mathematician sought to find the form, Of words and phrases, sentences and more, To craft a tale that readers would adore.

With numbers and equations, she did seek, To find the patterns hidden in the text. From words to sentences, her mind did peek, And found the structure that would be the next.

In paragraphs, she saw the chapters grow, Each sentence building on the one before. With math's precision, narratives did flow, Creating tales that readers would adore.

And so she crafted stories, pure and bright, With math and words, a beacon in the night.

Epic Poem 2: The Quest for Love

In the realm of hearts, where love does dwell, A mathematician sought to break the spell, Of loneliness and sorrow, hearts apart, And find the formula to heal the heart.

With equations, she did measure every beat, Predicting joy and sorrow, highs and lows. With every step, a new equation's feat, And love's true path, through math, it clearly shows.

But when love faltered, math did guide her still, To find the strength to leave, to start anew. With numbers, she did measure every thrill, And found the courage to bid love adieu.

And so she journeyed on, with math in hand, To find the love that she could understand.

Epic Poem 3: The Path of Transformation

In the land of change, where growth does bloom, A mathematician sought to find the room, For transformation, healing, and rebirth, And find the patterns hidden in the earth.

With every change, a new equation's start, And through its logic, she did find the way. From sorrow's depths, she rose with math's clear heart, To find the strength to heal, to grow, to stay.

In every change, a pattern she did see, A sequence leading to a brighter day. With math, she found the path to truly be, And in its guidance, she did find her way.

And so she journeyed on, with math as guide, To transformation's light, with hope and pride.

CHAPTER 5

Explain in detail, the literary criticism in the field of math, then show why, and offer a summary of excellent ideas to judge a math story, aka an expression.

Literary Criticism in the Field of Math

Literary criticism in the field of math involves analyzing and evaluating mathematical texts, stories, and expressions through the lens of literary theory. This interdisciplinary approach seeks to understand how mathematical concepts and structures can be used to create compelling narratives and how these narratives can be interpreted and appreciated. Let's explore this relationship in detail and provide some examples.

Relationship Between Math and Literary Criticism

1. Structure and Form: Mathematical texts and stories often have a clear structure and form, similar to literary works. Literary criticism can analyze the structure of mathematical narratives, examining how they are organized and how their form contributes to their meaning.

Example:
- **Math:** A mathematical proof follows a logical sequence of statements and conclusions.

- **Literary Criticism:** Analyzing the structure of a mathematical proof can reveal how the logical progression of ideas creates a coherent and persuasive argument.

2. Symbolism and Metaphor: Mathematical concepts and symbols can be used metaphorically in literature to convey deeper meanings. Literary criticism can explore how these symbols and metaphors enhance the narrative and contribute to its themes.

Example:
- **Math:** The concept of infinity can symbolize boundless possibilities or the unknown.
- **Literary Criticism:** Examining how the symbol of infinity is used in a story can reveal insights into the characters' aspirations and fears.

3. Patterns and Themes: Mathematical patterns and themes can be found in literary works, and literary criticism can analyze how these patterns contribute to the overall narrative. This includes examining recurring motifs, sequences, and structures.

Example:
- **Math:** The Fibonacci sequence is a pattern found in nature and art.

- **Literary Criticism:** Analyzing how the Fibonacci sequence is used in a story can reveal how the narrative mirrors natural growth and progression.

4. Language and Expression: Mathematical language and expressions can be analyzed for their aesthetic and rhetorical qualities. Literary criticism can explore how mathematical language is used to create meaning and evoke emotions.

Example:
- **Math:** The elegance of a mathematical equation can be appreciated for its simplicity and beauty.
- **Literary Criticism:** Examining the language of a mathematical equation in a story can reveal how it contributes to the narrative's aesthetic appeal.

Summary of Excellent Ideas to Judge a Math Story

1. **Clarity and Coherence:**
   - Evaluate how clearly and coherently the mathematical concepts are presented in the story. A well-structured narrative should guide the reader through the mathematical ideas without confusion.

2. **Symbolism and Metaphor:**
   - Analyze how mathematical symbols and metaphors are used to enhance the narrative.

Consider how these elements contribute to the story's themes and deeper meanings.

3. **Patterns and Themes:**
   - Examine the use of mathematical patterns and themes in the story. Look for recurring motifs and sequences that contribute to the overall narrative structure.

4. **Language and Expression:**
   - Assess the aesthetic and rhetorical qualities of the mathematical language used in the story. Consider how the language creates meaning and evokes emotions.

5. **Integration of Math and Narrative:**
   - Evaluate how seamlessly the mathematical concepts are integrated into the narrative. A successful math story should blend mathematical ideas with the plot and characters in a way that feels natural and engaging.

Sonnets

Sonnet 1: The Elegance of Math

In numbers' dance, a story comes to light, With symbols, patterns, themes that intertwine. The elegance of math, a pure delight, In prose and verse, its beauty we define.

A proof's clear structure, logic's steady hand, Reveals the path to truth with each new line. Infinity, a symbol vast and grand, In stories' depths, its endless bounds we find.

The Fibonacci's sequence, nature's art, In tales of growth and change, it finds its place. With language clear, expressions from the heart, Math's beauty shines, its elegance and grace.

In stories' weave, math's essence we embrace, A union rich, where art and logic trace.

Sonnet 2: The Patterns of Infinity

In tales of old, where symbols speak in code, The patterns of infinity unfold. With metaphors, the stories' paths are showed, In math's embrace, their secrets we behold.

A sequence grows, like nature's endless flow, In Fibonacci's steps, the tale is spun. With each new term, the narrative does grow, A journey vast, where math and art are one.

The language of equations, pure and bright, In stories' weave, its elegance we see. With clarity, it guides us through the night, A beacon true, in prose and poetry.

In math's embrace, the stories' truths we find, A union rich, where art and logic bind.

Epic Poems

Epic Poem 1: The Quest for Truth

In ancient lands, where stories weave their spell, A mathematician sought the truth to tell. With symbols bright and patterns pure and clear, He journeyed forth, with courage and no fear.

Through forests deep and mountains high he roamed, In search of wisdom, where the numbers loomed. With each new step, a sequence he did trace, The Fibonacci's path, a guiding grace.

In tales of old, the symbols spoke in code, Infinity's embrace, the stories showed. With metaphors, the truths began to shine, In math's embrace, the narrative divine.

The language of equations, pure and bright, In stories' weave, its elegance took flight. With clarity, it guided through the night, A beacon true, in prose and poetry's light.

In math's embrace, the stories' truths we find, A union rich, where art and logic bind.

Epic Poem 2: The Patterns of the Infinite

In lands of lore, where stories' paths are told, A mathematician sought the patterns bold. With symbols bright and sequences so grand, He journeyed forth, with courage in his hand.

Through deserts vast and oceans deep he sailed, In search of wisdom, where the numbers hailed. With each new term, a sequence he did trace, The Fibonacci's steps, a guiding grace.

In tales of old, the symbols spoke in code, Infinity's embrace, the stories showed. With metaphors, the truths began to shine, In math's embrace, the narrative divine. The language of equations, pure and bright, In stories' weave, its elegance took flight. With clarity, it guided through the night, A beacon true, in prose and poetry's light.

In math's embrace, the stories' truths we find, A union rich, where …

CHAPTER 6

Math and storytelling may seem like two entirely different realms?

And, they share a deep and intricate relationship. Both are systems of symbols and rules used to convey meaning and structure.

***

So, if you like, skip this quintessential function series ... and spark a new kernel.

Or offer proof of concept, such as dive into: how math relates to storytelling and explore some detailed examples.

***

Structure and Patterns

1. Plot Structure: Mathematical concepts like sequences and patterns can be applied to plot structures. For instance, the classic three-act structure (setup, confrontation, resolution) can be seen as a sequence of events that follow a specific pattern. This structure ensures a balanced and engaging narrative.

Example: In a mystery novel, the setup introduces the crime, the confrontation involves the investigation, and the resolution reveals the culprit. This sequence creates a logical flow that keeps readers engaged.

2. Symmetry and Balance: Symmetry in math refers to balanced proportions, which can be applied to storytelling to create harmony and coherence. Symmetrical structures in stories can make them more aesthetically pleasing and easier to follow.

Example: In "Pride and Prejudice" by Jane Austen, the symmetrical relationships between characters (e.g., Elizabeth and Darcy, Jane and Bingley) create a balanced narrative that explores themes of love and social class.

Character Development

3. Character Arcs: Mathematical functions can model character development. A character's growth can be represented as a curve on a graph, showing their progression over time. This helps in planning and visualizing the character's journey.

Example: In "Harry Potter," Harry's character arc can be seen as a positive slope, starting from an ordinary boy to becoming a hero. This progression follows a logical pattern of growth and development.

Themes and Symbolism

4. Symbolic Representation: Math uses symbols to represent concepts, and storytelling often employs symbolism to convey deeper meanings. Mathematical symbols can be used metaphorically in literature to enhance themes and messages.

Example: In "The Great Gatsby," the green light symbolizes Gatsby's unattainable dreams. This can be compared to a mathematical limit, representing something that is approached but never reached.

Rhythm and Pacing

5. Pacing and Timing: Mathematical concepts like ratios and proportions can be applied to the pacing of a story. Proper timing and rhythm ensure that the narrative flows smoothly and maintains the reader's interest.

Example: In "The Lord of the Rings," the pacing of the journey is carefully balanced with moments of action and rest, creating a rhythm that keeps readers engaged throughout the epic tale.

Detailed Examples

Example 1: Fibonacci Sequence in Storytelling The Fibonacci sequence (0, 1, 1, 2, 3, 5, 8, 13, ...) is a series of numbers where each number is the sum of the two preceding ones. This sequence can be applied to the structure of a story, where each chapter or section builds upon the previous ones, creating a natural and engaging progression.

Example 2: Golden Ratio in Visual Storytelling The golden ratio (approximately 1.618) is a mathematical ratio often found in nature and art. In visual storytelling, this ratio can be used to compose scenes and frames, creating aesthetically pleasing and harmonious visuals.

Example 3: Fractals in Narrative Complexity Fractals are complex patterns that are self-similar across different scales. In storytelling, fractal structures can be used to create intricate and layered narratives, where subplots mirror the main plot, adding depth and complexity.

Conclusion

Math.

CHAPTER 7

Math and phonemes, the smallest units of sound in a language, share a fascinating relationship rooted in patterns, structures, and probabilities.

If so, explore this relationship in detail and provide some examples.

Structure and Patterns

1. Phoneme Distribution: Mathematical principles can be used to analyze the distribution of phonemes in a language. By studying the frequency and patterns of phoneme usage, linguists can gain insights into the phonological structure of a language.

Example: In English, the phoneme /t/ occurs more frequently than the phoneme /ʒ/. By analyzing the distribution of these phonemes, linguists can identify common patterns and trends in spoken language.

Probabilities and Statistics

2. Phoneme Probabilities: Mathematical probabilities can be applied to predict the likelihood of certain phonemes occurring in specific contexts. This is particularly useful in speech recognition and synthesis technologies.

Example: In English, the phoneme /s/ is more likely to occur at the beginning of a word (e.g., "sun") than at the end (e.g., "bus"). By calculating the probabilities of phoneme positions, speech recognition systems can improve their accuracy.

CHAPTER 8

The relationship between math and intonation deploys from a shared foundation that bud patterns, rhythms, and branch treelike structures.

Aka, intonation, the rise and fall of pitch in speech maneuvers follows mathematical principles to reveal meaning, emphasize expressive words to reveal a spark, ideal, and convey situation emotion it that story stream series of events?

Or state another way:

Enhanced Statement: The relationship between math and intonation emerges from a shared foundation of patterns, rhythms, and branching tree-like structures. Intonation—the rise and fall of pitch in speech—follows mathematical principles to reveal meaning, emphasize expressive words, and convey emotions. This dynamic creates a narrative flow within a series of events, enhancing the story's impact.

Colorful Example: Imagine a symphony orchestra tuning up before a concert. Each instrument, from the deep rumble of the double bass to the high trill of the flute, adjusts to find harmony. Similarly, in a conversation, the rise and fall of our voice (intonation) adjusts to convey emotion and meaning. Just as musicians use mathematical principles to create a harmonious symphony, our intonation uses patterns and rhythms to make our speech more expressive. This mathematical underpinning helps us emphasize important words, reveal sparks of ideas, and paint vivid emotional landscapes within our stories.

Examples:

1. **Pitch Contours and Sine Waves**: The pitch contour of a sentence can be modeled using sine waves. For instance, in a question, the pitch tends to rise towards the end, resembling a sine wave pattern. In a statement, the pitch often falls, also following a sine-like pattern but in the opposite direction.

2. **Fourier Analysis**: Fourier analysis breaks down complex sound waves into simpler components, much like how you might separate a chord into individual notes. This is crucial in speech recognition technology, where understanding intonation can help machines comprehend context and meaning. Imagine dissecting the melody in a song to understand each instrument's contribution.

3. **Prosody and Probability**: Prosody, which includes intonation, can be analyzed using probability theory. For instance, in English, it's probable that a speaker will stress certain syllables in a sentence to convey meaning. "I didn't steal your pen" versus "I didn't steal your pen" has different intonation patterns highlighting different words, following predictable prosodic rules.

4. **Mathematical Models in Linguistics**: Phonologists use mathematical models to predict how intonation might change in different dialects or languages. For example, in tonal languages like Mandarin, the pitch of a word can change its meaning entirely. By mapping these tonal changes mathematically, linguists can study how intonation affects communication.

5. By seeing intonation through the lens of math, we grasp the deeper structure behind the melody of speech, revealing the harmony between numbers and sound.

6. **Or say another way:**

Intonation, the variation of pitch in spoken language, plays a crucial role in conveying meaning, emotion, and emphasis. The relationship between math and intonation lies in the mathematical principles that govern the patterns and structures of pitch variations. Let's explore this relationship in detail and provide some examples.

Mathematical Principles in Intonation

1. Frequency and Pitch: Pitch is determined by the frequency of sound waves. Higher frequencies correspond to higher pitches, and lower frequencies correspond to lower pitches. The relationship between frequency and pitch can be described mathematically using logarithmic scales, such as the musical octave, where each octave represents a doubling of frequency.

Example:

- **Math:** If a note has a frequency of 440 Hz (A4), the note an octave higher (A5) has a frequency of 880 Hz.

- **Intonation:** In speech, raising the pitch by an octave can indicate excitement or emphasis.

2. Waveforms and Harmonics: Intonation patterns can be analyzed using waveforms and harmonics. The shape and complexity of waveforms determine the quality of sound, while harmonics add richness and depth to the pitch.

Fourier analysis, a mathematical technique, can decompose complex waveforms into their constituent frequencies.

Example:

- **Math:** Fourier analysis can break down a complex waveform into a series of sine and cosine functions.
- **Intonation:** Analyzing the harmonic content of a speaker's voice can reveal subtle variations in pitch and tone that convey different emotions.

3. Prosody and Patterns: Prosody refers to the rhythm, stress, and intonation patterns in speech. These patterns can be described mathematically using concepts like amplitude, duration, and frequency. Prosodic features help distinguish between statements, questions, and commands.

Example:

- **Math:** A rising intonation pattern at the end of a sentence can be represented by an increase in frequency and amplitude.
- **Intonation:** "You're coming with us?" (rising intonation indicating a question) vs. "You're coming with us." (falling intonation indicating a statement).

Detailed Examples

Example 1: Pitch Contours in Speech Pitch contours represent the variation of pitch over time in spoken language. These contours can be plotted on a graph, with time on the x-axis and pitch (frequency) on the y-axis. Mathematical functions can model these contours to analyze and synthesize intonation patterns.

Example:

- **Math:** A quadratic function $y=ax^2+bx+c$ can model a pitch contour that rises and then falls.
- **Intonation:** "I can't believe it!" (rising and falling pitch indicating surprise).

Example 2: Stress and Emphasis Stress and emphasis in speech can be quantified using mathematical measures of amplitude and duration. Stressed syllables typically have higher amplitude and longer duration compared to unstressed syllables.

Example:

- **Math:** Amplitude A and duration D can be used to calculate the energy E of a syllable: $E = A \times D$.
- **Intonation:** "I didn't say he stole the money." (emphasis on different words changes the meaning).

- **Example 3: Melody and Speech** The melody of speech, or speech melody, can be analyzed using musical intervals and scales. Mathematical relationships between notes in a scale can be applied to the pitch variations in speech to study intonation patterns.

Example:

- **Math:** The interval between two notes can be calculated using the ratio of their frequencies.
- **Intonation:** "Hello!" (a melodic greeting with a specific pitch pattern).

Conclusion

Math and intonation are deeply interconnected, with mathematical principles providing a framework for understanding and analyzing pitch variations in speech. By applying these principles, linguists and speech scientists can gain insights into the patterns and structures of intonation, enhancing our understanding of spoken language. Whether it's through frequency, waveforms, prosody, or pitch contours, math offers valuable tools for studying and synthesizing?

CHAPTER 9

Stress in linguistics refers to the emphasis placed on certain syllables or words within a sentence. This emphasis is often described using mathematical concepts like patterns and probability.

Detailed Explanation: Stress follows rhythmic patterns, much like beats in music. These patterns can be modeled mathematically. For example, in English, the stress-timed rhythm means stressed syllables occur at regular intervals, creating a predictable pattern that can be measured and analyzed.

Examples:

1. **Stress Patterns in Poetry**: In a line of iambic pentameter, the stress pattern alternates, creating a rhythm of unstressed and stressed syllables (e.g., "Shall I compare thee to a summer's day?"). Mathematically, this can be represented as a binary sequence (010101), where 0 represents unstressed and 1 represents stressed syllables.

2. **Word Stress in Different Languages**: In English, word stress can change meaning. For instance,

"REcord" (noun) and "reCORD" (verb). Mathematically, the placement of stress can be analyzed using probability to predict which syllable is more likely to be stressed based on word position and phonological rules.

And by understanding stress through these patterns and probabilities, we grasp how emphasis shapes the meaning and rhythm of language.

Because the relationship between math and stress in language, specifically the emphasis placed on certain syllables or words, is rooted in the mathematical principles that govern patterns, rhythms, and structures in speech. So, let's change the channel or explore this relationship in detail and provide some examples.

Mathematical Principles in Stress

1. Amplitude and Intensity: Stress in language often involves an increase in amplitude (loudness) and intensity. These changes can be measured mathematically using decibels (dB). The amplitude of a stressed syllable is typically higher than that of an unstressed syllable.

Example:

- **Math:** If the amplitude of an unstressed syllable is 60 dB, a stressed syllable might have an amplitude of 70 dB.

- **Stress:** In the word "record," the stress on the first syllable (REcord) indicates a noun, while stress on the second syllable (reCORD) indicates a verb.

2. Duration and Timing: Stressed syllables often have a longer duration compared to unstressed syllables. This can be measured in milliseconds (ms) and analyzed using mathematical ratios.

Example:

- **Math:** If an unstressed syllable lasts 100 ms, a stressed syllable might last 150 ms.

- **Stress:** In the phrase "blackboard," the stress on the first syllable (BLACKboard) makes it a compound noun, while equal stress on both syllables (black BOARD) can imply a descriptive phrase.

3. Frequency and Pitch: Stress can also involve changes in pitch (frequency). Stressed syllables often have a higher pitch compared to unstressed syllables. This can be measured in Hertz (Hz).

Example:

- **Math:** If the pitch of an unstressed syllable is 200 Hz, a stressed syllable might have a pitch of 250 Hz.
- **Stress:** In the sentence "I didn't say he stole the money," stressing different words changes the meaning: "I didn't say he stole the money" vs. "I didn't say he stole the money."

Detailed Examples

Example 1: Stress Patterns in Poetry Poetry often uses specific stress patterns, such as iambic pentameter, where each line consists of five pairs of syllables, with the second syllable in each pair being stressed. This pattern can be represented mathematically as alternating 0s and 1s, where 0 represents an unstressed syllable and 1 represents a stressed syllable.

Example:

- **Math:** 0101010101
- **Stress:** "Shall I compare thee to a summer's day?" (Shall I comPARE thee TO a SUMmer's DAY?)

Example 2: Stress in Compound Words In compound words, the primary stress is usually on the first element. This can be analyzed using mathematical ratios to compare the amplitude and duration of the stressed and unstressed syllables.

Example:

- **Math:** Ratio of stressed to unstressed syllable amplitude = 1.5:1
- **Stress:** "GREENhouse" (a building for plants) vs. "green HOUSE" (a house that is green).

Example 3: Stress in Sentence Intonation In sentences, stress can be used to highlight important information or contrast. This can be represented using mathematical functions to model the changes in pitch and amplitude.

Example:

- **Math:** Function $f(x)=ax+b$ to model pitch change.
- **Stress:** "She didn't finish her homework" (emphasis on "didn't") vs. "She didn't finish her homework" (emphasis on "homework").

Conclusion

Math and stress in language are deeply interconnected, with mathematical principles providing a framework for understanding and analyzing the emphasis placed on certain syllables or words. By applying these principles, linguists and speech scientists can gain insights into the patterns and structures of stress, enhancing our understanding of spoken language. Whether it's through amplitude, duration, frequency, or stress patterns, math offers valuable tools for studying and synthesizing stress in language.

CHAPTER 10

The relationship between math and storytelling morphology sets in the structured, rule-based systems they both utilize. Morphology, the study of the structure of words, can be analyzed using mathematical concepts like sets, functions, and probabilities. Here's a detailed look:

Detailed Explanation:

1. **Pattern Recognition**: Just as math identifies patterns and structures, morphology identifies the patterns in word formation. For instance, the way prefixes, suffixes, and roots combine to form words follows specific rules and can be represented using algebraic models.

2. **Set Theory**: In math, set theory helps categorize elements. Similarly, in morphology, words are broken down into sets of morphemes (the smallest meaning-carrying units). For example, the word "unhappiness" consists of three morphemes: "un-," "happy," and "-ness." Each of these morphemes can be considered as elements in a set.

3. **Functions and Transformations**: Math uses functions to describe relationships between

elements. Morphology uses rules to transform base words into new forms. For example, adding "-ed" to a verb base indicates past tense (e.g., "walk" becomes "walked"). This transformation can be described as a function f(x) where f(walk)=walked.

4. **Probability and Statistics**: Probabilistic models in math can predict morphological changes. For example, the likelihood of using a regular versus irregular verb form can be analyzed statistically, helping understand language evolution and usage patterns.

Examples:

1. **Inflectional Morphology**: In English, verbs change form based on tense, number, and aspect. For instance, the verb "to run" changes to "ran" (past tense) and "running" (present participle). These changes follow predictable patterns that can be mathematically modeled.

2. **Derivational Morphology**: This involves creating new words by adding prefixes or suffixes. For example, adding "-er" to "teach" forms "teacher." This transformation can be viewed as a mathematical function where f(teach)=teacher.

3. **Allomorphy**: Some morphemes have different forms (allomorphs) depending on the context. For instance, the plural morpheme in English can be "-s," "-es," or change in irregular forms (e.g., "child" to "children"). These variations can be statistically analyzed to understand their distribution and usage.
4. **Context-Free Grammar**: Linguistics uses context-free grammars to describe the syntax and morphology of languages. This approach is akin to how mathematical logic uses production rules to describe formal languages. For example, the rule S→NP VP (a sentence is a noun phrase followed by a verb phrase) helps in parsing sentences in natural language processing.

Or say another way: Morphology in storytelling refers to the structure and formation of words, which is crucial for creating meaningful and engaging narratives. Math and morphology share a deep relationship through patterns, structures, and rules that govern the formation and transformation of words.

And if interested, explore this relationship in detail and notice these examples.

Mathematical Principles in Morphology

1. Patterns and Sequences: Mathematical patterns and sequences can be applied to the formation of words. For example, the use of prefixes, suffixes, and roots follows specific patterns that can be analyzed mathematically.

Example:
- **Math:** Arithmetic sequence $a, a+d, a+2d, a+3d, \ldots$
- **Morphology:** The word "unhappiness" can be broken down into the prefix "un-", the root "happy," and the suffix "-ness." The pattern of adding prefixes and suffixes follows a specific sequence.
- **2. Combinatorics:** Combinatorics, a branch of mathematics, deals with counting and arranging elements. In morphology, combinatorics can be used to analyze the possible combinations of morphemes (the smallest units of meaning) to form new words.

Example:
- **Math:** The number of combinations of n elements taken k at a time is given by (nk).
- **Morphology:** Given the morphemes "re-", "act," and "-ion," we can form the word "reaction." The possible combinations of these morphemes can be analyzed using combinatorial principles.

3. Algebraic Structures: Algebraic structures, such as groups and rings, aka groupoids, the algebraic structure system that consists of nonempty sets, underlying set, carrier set or domain, aka groupoids, monoid complexes and others, such as a picture can be worth a thousand words and reveal a story? And can be used to model the rules and transformations in morphology. These structures help in understanding how words change form based on grammatical rules.

Example:

- **Math:** A group is a set with an operation that combines any two elements to form a third element, satisfying certain conditions.
- **Morphology:** The transformation of the verb "run" to its past tense "ran" can be modeled using algebraic structures, where the operation represents the grammatical rule for forming past tense.
- **Detailed Examples**

Example 1: Inflectional Morphology Inflectional morphology involves changing the form of a word to express different grammatical categories, such as tense, number, or case. Mathematical functions can model these transformations.

Example:

- **Math:** Function f(x)=x+d to represent adding a suffix.
- **Morphology:** The word "walk" can be transformed to "walked" by adding the suffix "-ed" to indicate past tense.

Example 2: Derivational Morphology Derivational morphology involves creating new words by adding prefixes or suffixes to a base word. This process can be analyzed using mathematical sequences and patterns.

Example:

- **Math:** Sequence a,a+d,a+2d,... to represent adding morphemes.
- **Morphology:** The word "happiness" is derived from the base word "happy" by adding the suffix "-ness."

Example 3: Morphological Parsing Morphological parsing involves breaking down a word into its constituent morphemes. This process can be represented using tree structures and graphs, which are mathematical tools for analyzing hierarchical relationships.

Example:

- **Math:** Tree structure to represent hierarchical relationships.

- **Morphology:** The word "unbelievable" can be parsed into the morphemes "un-", "believe," and "-able," with a tree structure showing the relationships between these morphemes.

Conclusion

Math and morphology in storytelling are deeply interconnected, with mathematical principles providing a framework for understanding and analyzing the structure and formation of words. By applying these principles, linguists and writers can gain insights into the patterns and rules that govern word formation, enhancing our understanding of language and storytelling. Whether it's through patterns, sequences, combinatorics, or algebraic structures, math offers valuable tools for studying and synthesizing morphology in storytelling.

CHAPTER 11

Imagine **mathematics** and **prose syntax** as two master chefs in a bustling kitchen. At first glance, their culinary creations might seem entirely different, but when you look closer, you see they share a deep, underlying artistry.

Mathematics is like the precise, measured techniques of a pastry chef. Each number, equation, and theorem are an ingredient that must be carefully weighed and combined to create a perfect dessert. For instance, consider the elegance of a quadratic equation,

$y = x2 + 3x - 7.$

And this equation can be visualized as a beautifully layered cake, where each layer represents a different component of the equation, coming together to form a delicious whole.

Prose syntax, on the other hand, is akin to the creative flair of a gourmet chef. Sentences flow with flavor and texture, each word carefully chosen to enhance the overall dish. Think of a sentence like, "The apple was eaten." In linguistic terms, this simple sentence is a complex recipe, where each word plays a crucial role in the final taste.

Now, let's bring these two chefs together. In both mathematics and prose, structure is key. Just as a chef must understand the recipe to create a masterpiece, a writer must grasp the rules of syntax to craft compelling sentences. Similarly, a mathematician must follow logical steps to solve equations.

If you like, consider Noam Chomsky's linguistic theories, which break down sentences into tree-like diagrams. These diagrams show how our brains combine words and phrases to form coherent sentences, much like how chefs combine ingredients to create new dishes. This combining process in linguistics is akin to mixing mathematical functions to create new equations.

In essence, both mathematics and prose syntax are about **building and understanding structures**. They teach us to see the world in patterns, whether through numbers or flavors, enhancing our appreciation of the beauty and complexity in both fields.

: The Algebra of Language — Caltech Magazine

And it seems quite fascinating how these two seemingly different disciplines can come together to create something so harmonious and delightful?

Or says another way:

The Relationship Between Math and Prose Syntax

1. **Structure and Patterns**:
   - **Mathematics**: At its core, math is about finding patterns and structures. Equations, formulas, and geometric shapes all follow specific rules and structures.
   - **Prose Syntax**: Similarly, prose syntax involves the arrangement of words and sentences to create meaning. Just like a mathematical formula, a well-constructed sentence follows grammatical rules and patterns.

2. **Economy and Precision**:
   - **Mathematics**: Math requires precision and economy. Every symbol and number have a specific purpose and place.
   - **Prose Syntax**: Good prose also values precision. Each word should serve a purpose, contributing to the overall clarity and impact of the text.

3. **Logical Flow**:

- **Mathematics**: Mathematical proofs and equations follow a logical progression from premises to conclusion.
- **Prose Syntax**: Similarly, a well-written paragraph or essay follows a logical flow, guiding the reader from one idea to the next seamlessly.

Detailed Examples

1. **Mathematical Patterns in Poetry**:
    - **Example**: The Fibonacci sequence, where each number is the sum of the two preceding ones (1, 1, 2, 3, 5, 8…), can be used to structure a poem. Each line of the poem can have a number of syllables corresponding to the Fibonacci sequence.
    - **Poem**:
    - One
    - Small,
    - Precise,
    - Poetic,
    - Spirals of nature,

- Fibonacci's sequence unfolds.

2. **Logical Flow in Prose**:
    - **Example**: Consider a mathematical proof that follows a logical sequence of steps to reach a conclusion. Similarly, a persuasive essay might start with an introduction, followed by arguments supported by evidence, and conclude with a summary.
    - **Essay Structure**:
    - Introduction: Present the main argument.
    - Body Paragraph 1: First supporting point with evidence.
    - Body Paragraph 2: Second supporting point with evidence.
    - Body Paragraph 3: Third supporting point with evidence.
    - Conclusion: Summarize the argument and restate the main point.

3. **Economy and Precision in Writing**:
    - **Example**: In mathematics, the equation ($E = mc^2$) succinctly expresses the

relationship between energy, mass, and the speed of light. In prose, a sentence like "She smiled, a fleeting glimpse of joy" uses precise language to convey a vivid image.

- **Sentence**:
- "The sun set, painting the sky with hues of orange and pink."

And by understanding these connections, we can see how the precision and structure of mathematics can inform and enhance the art of writing prose. Both disciplines, though different in their methods, strive for clarity, beauty, and a deeper understanding of the world.

Or say another way:

Math and prose syntax might seem like two distinct realms, but they share a fascinating relationship rooted in structure, rules, and clarity. Both systems aim to convey meaning effectively, whether through numbers and symbols or words and sentences. Let's explore this relationship in detail and provide some examples.

Structure and Syntax

1. Syntax and Grammar: Just as math has rules and syntax for forming equations, prose has grammar rules for constructing sentences. These rules ensure that sentences are coherent and convey the intended meaning. For example, in English, a basic sentence structure follows the Subject-Verb-Object (SVO) pattern.

Example:

Math: $a+b=c$

- **Sentence:** "The cat (Subject) chased (Verb) the mouse (Object)."

Patterns and Sequences

2. Patterns in Language: Mathematical patterns can be found in the rhythm and flow of prose. For instance, the use of parallelism in writing creates a pattern that enhances readability and aesthetic appeal.

Example:

- **Math:** Arithmetic sequence $a, a+d, a+2d, a+3d, \ldots$
- **Sentence:** "She likes reading, writing, and drawing."

Balance and Symmetry

3. Symmetry in Sentences: Symmetry in math refers to balanced proportions, which can be applied to sentences to create harmony and coherence. Balanced sentences often use parallel structure to achieve this effect.

Example:

- **Math:** Symmetrical equation x=y
- **Sentence:** "To err is human; to forgive, divine."

Ratios and Proportions

4. Proportions in Language: Ratios and proportions in math can be applied to the length and complexity of sentences. Varying sentence lengths can create a pleasing rhythm and maintain the reader's interest.

Example:

- **Math:** Ratio ab
- **Sentence:** "He ran. She walked. They strolled together."

Detailed Examples

Example 1: Fibonacci Sequence in Sentences The Fibonacci sequence (0, 1, 1, 2, 3, 5, 8, 13, ...) is a series of numbers where each number is the sum of the two preceding ones. This sequence can be applied to the length of sentences in a paragraph to create a natural and engaging progression.

Example 2: Golden Ratio in Sentence Structure The golden ratio (approximately 1.618) is a mathematical ratio often found in nature and art. In writing, this ratio can be used to determine the ideal length of sentences and paragraphs, creating aesthetically pleasing and harmonious text.

Example 3: Fractals in Narrative Complexity Fractals are complex patterns that are self-similar across different scales. In writing, fractal structures can be used to create intricate and layered sentences, where sub-clauses mirror the main clause, adding depth and complexity.

Conclusion

Math and prose syntax are deeply interconnected, with mathematical principles providing structure, balance, and depth to sentences. By understanding and applying these principles, writers can create more engaging and meaningful prose. Whether it's through syntax, patterns, symmetry, or proportions, math offers valuable tools for crafting compelling sentences.

CHAPTER 12

Mathematics, semantics, and storytelling may seem like distinct fields, but they share deep connections that can enrich our understanding of each, or change the channel or explore these relationships with some vivid examples.

The Relationship Between Math, Semantics, and Storytelling

1. **Structure and Meaning**:
    - **Mathematics**: Math provides a structured way to describe relationships and patterns. Equations and formulas are precise and convey specific meanings.
    - **Semantics**: Semantics is the study of meaning in language. It focuses on how words and sentences convey meaning.
    - **Storytelling**: Storytelling uses language to create narratives that convey deeper meanings and emotions. The structure of a story (beginning, middle, end) helps organize these meanings.

2. **Logical Frameworks**:
   - **Mathematics**: Math relies on logical frameworks to solve problems and prove theorems.
   - **Semantics**: Logical frameworks in semantics help us understand how different parts of a sentence contribute to its overall meaning.
   - **Storytelling**: Logical progression in storytelling ensures that the narrative makes sense and engages the audience.
3. **Symbolism and Abstraction**:
   - **Mathematics**: Math uses symbols to represent abstract concepts (e.g., numbers, variables).
   - **Semantics**: Words are symbols that represent objects, actions, and ideas.
   - **Storytelling**: Stories often use symbols and metaphors to convey deeper meanings and abstract concepts.

Detailed Examples

1. **Mathematical Patterns in Storytelling**:

- **Example**: The structure of a story can follow mathematical patterns. For instance, the "Hero's Journey" narrative structure can be mapped onto a circular pattern, reflecting the cyclical nature of many mathematical concepts.
- **Story Structure**:
- Departure: The hero leaves their ordinary world.
- Initiation: The hero faces challenges and gains new knowledge.
- Return: The hero returns transformed.

2. **Logical Frameworks in Semantics and Storytelling**:
    - **Example**: Consider a mathematical proof that follows a logical sequence of steps to reach a conclusion. Similarly, a well-constructed argument in a story or essay follows a logical progression.
    - **Story Example**:
    - Premise: The protagonist is unhappy with their life.

- Development: They embark on a journey to find happiness.
- Conclusion: They discover that happiness was within them all along.

3. **Symbolism and Abstraction in Math and Storytelling**:
    - **Example**: In mathematics, the equation ($E = mc^2$) symbolizes the relationship between energy and mass. In storytelling, a character's journey can symbolize broader human experiences.
    - **Story Example**:
    - The dragon in the story represents the protagonist's inner fears.
    - The treasure symbolizes self-discovery and personal growth.

By understanding these connections, we can see how the precision and structure of mathematics can inform and enhance the art of storytelling. Both disciplines, though different in their methods, strive for clarity, beauty, and a deeper understanding of the world.

Or say another way:

Math and semantics might seem like two distinct fields, but they share a deep and intricate relationship, especially in the context of storytelling. Semantics is the study of meaning in language, while math provides the tools and structures to analyze and model these meanings. Let's explore this relationship in detail and provide some examples.

Mathematical Principles in Semantics

1. Formal Semantics: Formal semantics uses mathematical logic to represent and analyze the meanings of sentences. This involves using symbols and formulas to capture the relationships between words and their meanings.

Example:

- **Math:** Predicate logic can represent the meaning of sentences. For instance, the sentence "All humans are mortal" can be represented as $\forall x(H(x) \rightarrow M(x))$, where $H(x)$ means "x is a human" and $M(x)$ means "x is mortal."

- **Semantics:** This formal representation helps in understanding the logical structure and implications of the sentence.

2. Set Theory: Set theory, a branch of mathematical logic, is used to model the meanings of words and phrases. Words can be seen as sets of entities, and their meanings can be analyzed through set operations like union, intersection, and complement.

Example:

- **Math:** The word "cat" can be represented as a set of all cats. The phrase "black cat" can be represented as the intersection of the set of all cats and the set of all black things.
- **Semantics:** This helps in understanding how meanings combine and interact in phrases and sentences.

3. Probability and Statistics: Probability and statistics are used to model the likelihood of different meanings and interpretations. This is particularly useful in natural language processing and machine learning, where algorithms need to determine the most likely meaning of a sentence based on context.

Example:

- **Math:** Bayesian inference can be used to update the probability of a meaning given new evidence. For instance, if the word "bank" appears in a sentence, the probability of it meaning "financial institution" vs. "riverbank" can be updated based on the surrounding words.

- **Semantics:** This helps in disambiguating words with multiple meanings and understanding context-dependent interpretations.

Detailed Examples

Example 1: Compositional Semantics Compositional semantics is the principle that the meaning of a sentence is determined by the meanings of its parts and the rules used to combine them. This can be modeled using functions and operations from math.

Example:

- **Math:** If f represents the meaning of a verb and g represents the meaning of a noun, the meaning of the verb-noun phrase can be represented as $f(g(x))$.

- **Semantics:** In the sentence "The cat sleeps," the meaning of "sleeps" (a function) is applied to the meaning of "the cat" (an entity), resulting in the overall meaning of the sentence.

Example 2: Vector Space Models Vector space models represent words and their meanings as vectors in a high-dimensional space. The relationships between words can be analyzed using vector operations like addition, subtraction, and dot product.

Example:

- **Math:** The word "king" can be represented as a vector $\vec{k}$, and the word "queen" can be represented as a vector $\vec{q}$. The relationship between "king" and "queen" can be analyzed using the vector difference $\vec{k}-\vec{q}$.

- **Semantics:** This helps in understanding semantic relationships and analogies, such as "king is to queen as man is to woman."

Example 3: Semantic Networks Semantic networks use graph theory to represent the relationships between words and concepts. Nodes represent words or concepts, and edges represent the relationships between them.

Example:

- **Math:** A graph $G=(V,E)$ can represent a semantic network, where V is the set of nodes (words) and E is the set of edges (relationships).

- **Semantics:** This helps in visualizing and analyzing the connections between different words and concepts, such as synonyms, antonyms, and hierarchical relationships.

Conclusion

Math and semantics in storytelling are deeply interconnected, with mathematical principles providing a framework for understanding and analyzing the meanings of words and sentences. By applying these principles, linguists and writers can gain insights into the patterns and structures that govern meaning, enhancing our understanding of language and storytelling. Whether it's through formal semantics, set theory, probability, or X.

CHAPTER 13

Mathematics, pragmatics, and storytelling might seem like separate realms, but they intertwine in fascinating ways that can deepen our understanding of each.

If so, let's explore these connections with some vivid examples.

The Relationship Between Math, Pragmatics, and Storytelling

1. **Context and Meaning**:
    - **Mathematics**: Math provides a structured way to describe relationships and patterns. Equations and formulas are precise and convey specific meanings.
    - **Pragmatics**: Pragmatics is the study of how context influences the interpretation of meaning in language. It looks at how people understand language in different situations.
    - **Storytelling**: Storytelling uses language to create narratives that convey deeper meanings and emotions. The context of a

story (setting, characters, plot) helps shape its meaning.

2. **Logical Frameworks**:
   - **Mathematics**: Math relies on logical frameworks to solve problems and prove theorems.
   - **Pragmatics**: Pragmatics uses logical frameworks to understand how language functions in different contexts, such as how a statement can imply something beyond its literal meaning.
   - **Storytelling**: Logical progression in storytelling ensures that the narrative makes sense and engages the audience.

3. **Symbolism and Abstraction**:
   - **Mathematics**: Math uses symbols to represent abstract concepts (e.g., numbers, variables).
   - **Pragmatics**: Words and phrases can have different meanings depending on the context, much like how symbols in math can represent different values or concepts.

- Storytelling: Stories often use symbols and metaphors to convey deeper meanings and abstract concepts.

Detailed Examples

1. **Mathematical Patterns in Storytelling**:
   - **Example**: The structure of a story can follow mathematical patterns. For instance, the "Hero's Journey" narrative structure can be mapped onto a circular pattern, reflecting the cyclical nature of many mathematical concepts.
   - **Story Structure**:
   - Departure: The hero leaves their ordinary world.
   - Initiation: The hero faces challenges and gains new knowledge.
   - Return: The hero returns transformed.

2. **Logical Frameworks in Pragmatics and Storytelling**:
   - **Example**: Consider a mathematical proof that follows a logical sequence of steps to reach a conclusion. Similarly, a well-

constructed argument in a story or essay follows a logical progression.

- **Story Example**:
- Premise: The protagonist is unhappy with their life.
- Development: They embark on a journey to find happiness.
- Conclusion: They discover that happiness was within them all along.

3. **Symbolism and Abstraction in Math and Storytelling**:

 - **Example**: In mathematics, the equation ($E = mc^2$) symbolizes the relationship between energy and mass. In storytelling, a character's journey can symbolize broader human experiences.
 - **Story Example**:
 - The dragon in the story represents the protagonist's inner fears.
 - The treasure symbolizes self-discovery and personal growth.

By understanding these connections, we can see how the precision and structure of mathematics can inform and enhance the art of storytelling. Both disciplines, though different in their methods, strive for clarity, beauty, and a deeper understanding of the world.

Or say another way:

Pragmatics is the branch of linguistics that deals with the use of language in context and the ways in which people produce and comprehend meanings through language. It involves understanding the intentions behind words, the context in which they are spoken, and the social rules governing communication. Math and pragmatics share a relationship through the application of mathematical principles to analyze and model these aspects of language use. And let's explore this relationship in detail and provide some examples.

Mathematical Principles in Pragmatics

1. Probability and Inference: Pragmatics often involves making inferences based on context and prior knowledge. Probability theory can be used to model these inferences, helping to predict the likelihood of certain interpretations or responses.

Example:

- **Math:** Bayes' Theorem can be used to update the probability of a hypothesis based on new evidence.
- **Pragmatics:** If someone says, "It's cold in here," the probability that they want the window closed increases if the context is a room with an open window.

2. Game Theory: Game theory, a branch of mathematics that studies strategic interactions, can be applied to pragmatics to model conversational dynamics. It helps in understanding how speakers choose their words and how listeners interpret them based on the expected responses.

Example:

- **Math:** The Nash Equilibrium is a concept in game theory where no player can benefit by changing their strategy while the other players keep theirs unchanged.
- **Pragmatics:** In a conversation, if both speakers choose their words to maximize mutual understanding and cooperation, they reach a conversational equilibrium.

3. Information Theory: Information theory deals with the quantification, storage, and communication of information.

It can be applied to pragmatics to analyze how efficiently information is conveyed in a conversation.

Example:

- **Math:** Shannon's Entropy measures the uncertainty or information content in a message.
- **Pragmatics:** A speaker's choice of words can minimize entropy by reducing ambiguity and ensuring clarity.

Detailed Examples

Example 1: Implicature and Probability Implicature refers to the implied meaning that is not explicitly stated. Probability theory can model the likelihood of different implicatures based on context.

Example:

- **Math:** Using conditional probability $P(A|B)$ to model the likelihood of an implicature given the context.
- **Pragmatics:** If someone says, "Can you pass the salt?" the probability that they are making a request rather than asking about your ability increases in the context of a dinner table.

Example 2: Politeness and Game Theory Politeness strategies in conversation can be analyzed using game theory to understand how speakers balance the need to be polite with the need to convey information.

Example:

- **Math:** The concept of a payoff matrix in game theory to model the costs and benefits of different politeness strategies.
- **Pragmatics:** Choosing to say, "Would you mind closing the window?" instead of "Close the window" involves a trade-off between politeness and directness.

Example 3: Context and Information Theory The context in which a conversation takes place can be analyzed using information theory to understand how it affects the efficiency of communication.

Example:

- **Math:** Calculating the mutual information between the context and the message to measure how much information the context provides about the message.
- **Pragmatics:** In a noisy environment, a speaker might choose simpler, more redundant language to

ensure the message is understood, thereby increasing mutual information.

Conclusion

Math and pragmatics are deeply interconnected, with mathematical principles providing a framework for understanding and analyzing the use of language in context. By applying these principles, linguists and communication theorists can gain insights into the patterns and structures of pragmatic language use, enhancing our understanding of communication. Whether it's through probability, game theory, or information theory, math offers valuable tools for studying and modeling pragmatics in a complete story.

CHAPTER 14

Mathematics and orthography (the conventional spelling system of a language) may seem like distant cousins, but they share intriguing connections that can illuminate both fields.

If so, let's explore these relationships with some vivid examples.

The Relationship Between Math and Orthography in Storytelling

1. **Structure and Rules**:
    - **Mathematics**: Math is built on a foundation of rules and structures. Equations, formulas, and geometric shapes all follow specific guidelines.
    - **Orthography**: Similarly, orthography relies on rules for spelling, punctuation, and grammar. These rules ensure that written language is clear and understandable.
    - **Storytelling**: In storytelling, both math and orthography contribute to the clarity and

coherence of the narrative. Proper spelling and punctuation help convey the story accurately, while mathematical structures can add layers of meaning.

2. **Patterns and Consistency**:
   - **Mathematics**: Math is all about recognizing and creating patterns. Whether it's the Fibonacci sequence or geometric symmetry, patterns are central to mathematical thinking.
   - **Orthography**: Spelling and grammar rules create patterns in language. Consistent use of these patterns helps readers understand and predict the flow of text.
   - **Storytelling**: In storytelling, patterns in language (like rhyme schemes in poetry or repetitive structures in prose) can enhance the narrative and make it more engaging.

3. **Symbolism and Abstraction**:
   - **Mathematics**: Math uses symbols (like numbers and variables) to represent abstract concepts.

- **Orthography**: Letters and punctuation marks are symbols that represent sounds and pauses in speech.
- **Storytelling**: Stories often use symbols and metaphors to convey deeper meanings. Mathematical symbols can be woven into narratives to add layers of abstraction and symbolism.

Detailed Examples

1. **Mathematical Patterns in Poetry**:
    - **Example**: The Fibonacci sequence can be used to structure a poem. Each line of the poem can have a number of syllables corresponding to the Fibonacci sequence.
    - **Poem**:
    - One
    - Small,
    - Precise,
    - Poetic,
    - Spirals of nature,
    - Fibonacci's sequence unfolds.

2. **Logical Frameworks in Storytelling**:
    - **Example**: Consider a mathematical proof that follows a logical sequence of steps to reach a conclusion. Similarly, a well-constructed story follows a logical progression.
    - **Story Example**:
    - Premise: The protagonist is unhappy with their life.
    - Development: They embark on a journey to find happiness.
    - Conclusion: They discover that happiness was within them all along.
3. **Symbolism and Abstraction in Math and Storytelling**:
    - **Example**: In mathematics, the equation ($E = mc^2$) symbolizes the relationship between energy and mass. In storytelling, a character's journey can symbolize broader human experiences.
    - **Story Example**:

- The dragon in the story represents the protagonist's inner fears.
- The treasure symbolizes self-discovery and personal growth.

By understanding these connections, we can see how the precision and structure of mathematics can inform and enhance the art of storytelling. Both disciplines, though different in their methods, strive for clarity, beauty, and a deeper understanding of the world.

Or say another way:

Orthography refers to the conventional spelling system of a language, including the rules for writing words correctly. The relationship between math and orthography lies in the mathematical principles that govern patterns, structures, and rules in spelling and writing. Let's explore this relationship in detail and provide some examples.

Mathematical Principles in Orthography

1. Patterns and Sequences: Mathematical patterns and sequences can be applied to the spelling of words. For example, the use of prefixes, suffixes, and roots follows specific patterns that can be analyzed mathematically.

Example:

- **Math:** Arithmetic sequence a,a+d,a+2d,a+3d,…
- **Orthography:** The word "unhappiness" can be broken down into the prefix "un-", the root "happy," and the suffix "-ness." The pattern of adding prefixes and suffixes follows a specific sequence.

2. Combinatorics: Combinatorics, a branch of mathematics, deals with counting and arranging elements. In orthography, combinatorics can be used to analyze the possible combinations of letters and morphemes to form new words.

Example:

- **Math:** The number of combinations of n elements taken k at a time is given by (nk).
- **Orthography:** Given the letters "c," "a," "t," we can form the word "cat." The possible combinations of these letters can be analyzed using combinatorial principles.

3. Probability and Frequency: Probability theory can be applied to orthography to analyze the frequency of letter combinations and predict the likelihood of certain spellings. This helps in understanding common spelling patterns and errors.

Example:

- **Math:** The probability P(A) of an event A occurring is given by P(A)=Number of favorable outcomes Total number of outcomes.

- **Orthography:** The probability of the letter "e" appearing at the end of a word in English is higher than that of the letter "z."

4. Graph Theory: Graph theory, a branch of mathematics that studies the relationships between objects, can be applied to orthography to analyze the connections between letters and sounds in words.

Example:

- **Math:** A graph consists of vertices (nodes) connected by edges (lines).

- **Orthography:** A graph can represent the connections between letters in a word, with vertices representing letters and edges representing the transitions between them.

Detailed Examples

Example 1: Spelling Patterns in English Spelling patterns in English often follow specific rules, such as the "i before e except after c" rule. These patterns can be analyzed using mathematical sequences and probabilities.

Example:

- **Math:** Sequence a,a+d,a+2d,… to represent adding letters.

- **Orthography:** The word "receive" follows the "i before e except after c" rule, while "believe" follows the "i before e" pattern.

Example 2: Morphological Parsing Morphological parsing involves breaking down a word into its constituent morphemes. This process can be represented using tree structures and graphs, which are mathematical tools for analyzing hierarchical relationships.

Example:

- **Math:** Tree structure to represent hierarchical relationships.

- **Orthography:** The word "unbelievable" can be parsed into the morphemes "un-", "believe," and "-able," with a tree structure showing the relationships between these morphemes.

Example 3: Frequency Analysis in Spelling Frequency analysis involves analyzing the frequency of letter combinations in a language. This can be done using probability theory to predict common spelling patterns and errors.

Example:

- **Math:** Calculating the frequency f of a letter combination x in a corpus of text.

- **Orthography:** The letter combination "th" is one of the most frequent in English, while "zx" is rare.

Conclusion

Math and orthography in storytelling are deeply interconnected, with mathematical principles providing a framework for understanding and analyzing the structure and rules of spelling. By applying these principles, linguists and writers can gain insights into the patterns and rules that govern orthography, enhancing our understanding of language and storytelling. Whether it's through patterns, sequences, combinatorics, probability, or graph theory, math offers valuable tools for studying and synthesizing orthography in storytelling.

CHAPTER 15

Imagine **mathematics** and the **Prosody storytelling system** as two master architects working together to design a magnificent building.

And at first glance, their tools and methods might seem different, but they share a deep, underlying connection that enhances the beauty and functionality of their creation.

Mathematics is like the precise blueprints and structural calculations that ensure the building stands tall and strong. Each number, equation, and theorem are an essential part of the design, providing the foundation and framework. For instance, consider the elegance of a quadratic equation,

$y = x^2 + 3x - 7$.

And this equation can be visualized as the blueprint for a parabolic arch, a key structural element that adds both strength and beauty to the building.

Prosody, on the other hand, is akin to the artistic flourishes and aesthetic details that bring the building to life. Prosody involves the rhythm, stress, and intonation of speech, much like how an architect uses curves, colors, and textures to create an inviting and engaging space. Think of a sentence like, "The apple was eaten." In prosodic terms, this simple sentence can be transformed by varying the pitch, stress, and rhythm to convey different emotions and meanings, much like how different architectural styles can evoke different feelings.

Now, let's bring these two architects together. In both mathematics and prosody, **structure** is key. Just as an architect must understand the principles of engineering to create a safe and functional building, a storyteller must grasp the rules of prosody to craft compelling narratives. Similarly, a mathematician must follow logical steps to solve equations, ensuring the integrity of their work.

Consider how prosody can be used to enhance the storytelling of mathematical concepts. For example, when explaining the Pythagorean theorem, a storyteller might use a rhythmic and engaging tone to emphasize the relationship between the sides of a right triangle: "In a right triangle, the square of the hypotenuse is equal to the sum of the squares of the other two sides." The prosodic elements—such as stressing key words and using a melodic intonation—help make the mathematical concept more memorable and easier to understand.

In essence, both mathematics and prosody are about **building and understanding structures**. They teach us to see the world in patterns, whether through numbers or sounds, enhancing our appreciation of the beauty and complexity in both fields.

Or says another way:

Prosody in storytelling refers to the patterns of rhythm, stress, and intonation in speech. It plays a crucial role in conveying meaning, emotion, and emphasis in both spoken and written narratives. The relationship between math and prosody lies in the mathematical principles that govern these patterns and structures.

And let's explore this relationship in detail and provide some examples.

Mathematical Principles in Prosody

1. Rhythm and Meter: Rhythm in prosody refers to the pattern of sounds and silences in speech. Meter, a structured rhythm in poetry and verse, can be analyzed using mathematical concepts such as sequences and periodic functions.

Example:

- **Math:** An arithmetic sequence a,a+d,a+2d,a+3d,… can represent a regular pattern of stressed and unstressed syllables.

- **Prosody:** In iambic pentameter, each line consists of five pairs of syllables, with the second syllable in each pair being stressed: "Shall I comPARE thee TO a SUMmer's DAY?"

2. Frequency and Pitch: Pitch is determined by the frequency of sound waves. Higher frequencies correspond to higher pitches, and lower frequencies correspond to lower pitches. The relationship between frequency and pitch can be described mathematically using logarithmic scales, such as the musical octave, where each octave represents a doubling of frequency.

Example:

- **Math:** If a note has a frequency of 440 Hz (A4), the note an octave higher (A5) has a frequency of 880 Hz.

- **Prosody:** In speech, raising the pitch by an octave can indicate excitement or emphasis.

3. Waveforms and Harmonics: Prosody patterns can be analyzed using waveforms and harmonics. The shape and complexity of waveforms determine the quality of sound, while harmonics add richness and depth to the pitch. Fourier analysis, a mathematical technique, can decompose complex waveforms into their constituent frequencies.

Example:

- **Math:** Fourier analysis can break down a complex waveform into a series of sine and cosine functions.

- **Prosody:** Analyzing the harmonic content of a speaker's voice can reveal subtle variations in pitch and tone that convey different emotions.

4. Stress and Emphasis: Stress in language involves an increase in amplitude (loudness) and intensity. These changes can be measured mathematically using decibels (dB). The amplitude of a stressed syllable is typically higher than that of an unstressed syllable.

Example:

- **Math:** If the amplitude of an unstressed syllable is 60 dB, a stressed syllable might have an amplitude of 70 dB.

- **Prosody:** In the word "record," the stress on the first syllable (REcord) indicates a noun, while stress on the second syllable (reCORD) indicates a verb.

Detailed Examples

Example 1: Pitch Contours in Speech Pitch contours represent the variation of pitch over time in spoken language. These contours can be plotted on a graph, with time on the x-axis and pitch (frequency) on the y-axis. Mathematical functions can model these contours to analyze and synthesize prosody patterns.

Example:

- **Math:** A quadratic function $y=ax^2+bx+c$ can model a pitch contour that rises and then falls.
- **Prosody:** "I can't believe it!" (rising and falling pitch indicating surprise).

Example 2: Stress Patterns in Poetry Poetry often uses specific stress patterns, such as iambic pentameter, where each line consists of five pairs of syllables, with the second syllable in each pair being stressed. This pattern can be represented mathematically as alternating 0s and 1s, where 0 represents an unstressed syllable and 1 represents a stressed syllable.

Example:

- **Math:** 0101010101

- **Prosody:** "Shall I compare thee to a summer's day?" (Shall I comPARE thee TO a SUMmer's DAY?)

**Example 3: Rhythm

CHAPTER 16

Let's go on a vivid, epic journey where mathematics and storytelling intertwine in various forms.

And we'll start with a complete story, then transform it into proofs, mathematical functions, epic poems, folk guitar songs, and popular TV episodes.

Complete Story: The Quest for the Golden Ratio

In a mystical land, there was a legendary architect named Pythagoras who sought the secret to perfect harmony in design. He believed that the key lay in a magical number known as the Golden Ratio. Pythagoras embarked on a quest, traveling through enchanted forests and across vast deserts, solving mathematical puzzles left by ancient sages.

One day, he encountered a wise old sage who presented him with a challenge: "To find the Golden Ratio, you must solve the riddle of the right triangle." Pythagoras pondered and realized that the answer lay in the relationship between the sides of the triangle. He discovered that in a right triangle, the square of the hypotenuse is equal to the sum of the squares of the other two sides. This revelation led him to the Golden Ratio, a number that appeared in the proportions of nature, art, and architecture.

With this newfound knowledge, Pythagoras returned to his homeland and built magnificent structures that stood the test of time, all based on the principles of the Golden Ratio. His legacy lived on, inspiring generations to explore the beauty of mathematics in the world around them.

Series of Proofs

1. **Pythagorean Theorem Proof:**
   - Given a right triangle with sides (a), (b), and hypotenuse (c):

   a2+b2=c2

2. **Golden Ratio Proof:**
   - The Golden Ratio, (ϕ), is defined as:

   ϕ=21+5

   - It satisfies the equation:

   ϕ2=ϕ+1

3. **Relationship in Nature:**

- The Fibonacci sequence, where each number is the sum of the two preceding ones, approximates the Golden Ratio as it progresses:

$$F(n)=F(n-1)+F(n-2)$$

Series of Math Functions

1. **Quadratic Function:**
    - The quadratic equation representing the Golden Ratio:

$$x^2-x-1=0$$

2. **Exponential Growth:**
    - Fibonacci sequence expressed as a function:

$$F(n)=5\phi n-(1-\phi)n$$

3. **Logarithmic Spiral:**
    - The logarithmic spiral found in nature, related to the Golden Ratio:

$$r=ae^{b\theta}$$

Series of Epic Poems

1. **The Architect's Dream:**
    - In ancient lands where wisdom flows, Pythagoras sought the secret that grows. Through trials and puzzles, he did roam, To find the ratio that calls nature home.

2. **The Sage's Challenge:**
    - "Solve the riddle of the right triangle," said the sage, "And unlock the harmony of every age." With squares and roots, the answer came, The Golden Ratio, forever to acclaim.

3. **The Legacy:**
    - With knowledge vast, he built with grace, Structures that time could not erase. The Golden Ratio, his guiding light, In every form, a wondrous sight.

Series of Folk Guitar Songs

1. **The Quest Begins:**
    - (Verse) In a land of ancient lore, Pythagoras set out to explore, With a heart full of dreams and a mind so bright, He sought the number that felt just right.

2. **The Triangle's Secret:**
   - (Chorus) Oh, the right triangle's secret, In the squares of sides, you'll meet it, The hypotenuse reveals the way, To the Golden Ratio's bright day.

3. **Building Wonders:**
   - (Bridge) With the ratio in his hand, He built wonders across the land, From nature's code to art's delight, The Golden Ratio shining bright.

Series of Popular TV Episodes

1. **Episode 1: The Architect's Journey**
   - Pythagoras sets out on his quest, encountering various mathematical challenges and mystical creatures.

2. **Episode 2: The Sage's Riddle**
   - Pythagoras meets the wise sage and solves the riddle of the right triangle, discovering the Golden Ratio.

3. **Episode 3: The Return Home**

- Armed with new knowledge, Pythagoras returns to his homeland and begins constructing magnificent buildings.

4. **Episode 4: The Legacy Lives On**
    - The impact of Pythagoras's discoveries on future generations and their applications in modern science and art.

Series Summary

In this series, we follow the journey of Pythagoras as he uncovers the secrets of the Golden Ratio. Through mathematical proofs, functions, epic poems, folk songs, and TV episodes, we explore the profound relationship between mathematics and storytelling, illustrating how these two disciplines can beautifully complement each other.

Isn't it amazing how mathematics can be expressed in so many vivid and engaging ways?

Or say another way:

Complete Story: The Math of Prosody in Storytelling

In the vibrant city of Numeria, a mathematician named Clara was fascinated by the relationship between math and prosody in storytelling. She believed that mathematical principles could unlock new insights into the rhythm, stress, and intonation patterns in speech. Clara began by analyzing classic poems and speeches, identifying patterns in their prosody.

She created a mathematical model to quantify these elements, assigning values to different aspects of prosody. Clara called it the Prosody Analysis Equation: $P=(R+S+I)T$, where P represented the prosody score, R represented rhythm, S represented stress, I represented intonation, and T represented time.

Clara tested her equation by applying it to various poems and speeches. She found that the equation accurately predicted the emotional impact and effectiveness of each piece. Encouraged by her success, Clara shared her findings with poets, writers, and public speakers. They began to use the Prosody Analysis Equation to guide their work, making adjustments to create more engaging and impactful performances.

The city of Numeria thrived, with artists producing works that resonated deeply with audiences. Clara's story proved that math could be a powerful tool in understanding and creating prosody in storytelling. Her innovative approach inspired others to embrace mathematical principles in their artistic pursuits. Numeria became a beacon of artistic excellence, demonstrating that a balanced life, guided by math, could lead to lasting artistic achievement.

Series of Proofs

1. **Proof of Rhythm Analysis:**
    - **Math:** Fourier analysis can decompose complex waveforms into their constituent frequencies.
    - **Prosody:** Analyzing the rhythmic patterns in a poem can reveal the underlying structure that contributes to its emotional impact.

2. **Proof of Stress Patterns:**
    - **Math:** Amplitude A and duration D can be used to calculate the energy E of a syllable: $E = A \times D$.
    - **Prosody:** Measuring the amplitude and duration of stressed syllables in a speech can quantify their emphasis and impact.

3. **Proof of Intonation Contours:**

- **Math:** A quadratic function y=ax2+bx+c can model a pitch contour that rises and then falls.

- **Prosody:** Plotting the pitch contour of a sentence can reveal the intonation pattern that conveys different emotions.

Series of Math Functions

1. **Rhythm Function:**

    - R(t)=∑n=1NAnsin⁡(2πfnt+ϕn)

    - Represents the rhythmic pattern of a poem or speech.

2. **Stress Function:**

    - S=∑i=1MAi×Di

    - Calculates the total stress energy in a piece of speech.

3. **Intonation Function:**

    - I(t)=at2+bt+c

    - Models the pitch contour of a sentence.

Series of Epic Poems

Epic Poem 1: The Quest for Prosody

In Numeria, a city bright and grand, A mathematician took a daring stand. Clara was her name, a mind so keen, She sought to find a way to intervene.

The search for rhythm that plagued her heart and mind, She sought to ease with formulas to find. She crafted tools to light the way, To balance stress, and pitch, and love each day.

The Prosody Analysis Equation was her key, To finding truth and living happily. Her town embraced her method, clear and bright, And found their tales were filled with pure delight.

Through trials and tests, they forged a path so bright, Their formula a beacon in the night. Clara's story spread, a legend far and wide, Of how math's power turned the prosody tide.

Epic Poem 2: Clara's Triumph

In Numeria, where stories come alive, A mathematician sought to help them thrive. Clara was her name, a mind so bright, She sought to bring the hidden truths to light.

With math and data, she did seek to find, A method to bring peace to troubled mind. She crafted tools to light the way, To balance stress, and pitch, and love each day.

The Prosody Analysis Equation was her key, To finding truth and living happily. Her town embraced her method, clear and bright, And found their tales were filled with pure delight.

Through trials and tests, they forged a path so bright, Their formula a beacon in the night. Clara's story spread, a legend far and wide, Of how math's power turned the prosody tide.

Series of Folk Guitar Songs

Folk Song 1: Clara's Equation

(Verse 1) In the town of Numeria, where stories come alive, A mathematician named Clara sought to help them thrive. With numbers and equations, she sought to find the way, To bring the hidden meanings to the light of day.

(Chorus) Oh, Clara's equation, lighting up the night, Bringing truth and wisdom, making stories bright. With math and love together, she found the perfect blend, Creating tales of wonder, that never seem to end.

(Verse 2) She analyzed the rhythm, the stress, and intonation, Creating a formula for every situation. Her Prosody Equation brought stories to new heights, Filling hearts with wonder, and minds with pure delight.

(Chorus) Oh, Clara's equation, lighting up the night,
Bringing truth and wisdom, making stories bright. With
math and love together, she found the perfect blend,
Creating tales of wonder, that never seem to end.

Series of Popular TV Episodes

Episode 1: The Prosody Equation Clara, a brilliant mathematician, discovers a way to analyze and enhance the prosody of stories using mathematical principles. She shares her findings with poets and writers, transforming the way they create and perform their works.

Episode 2: The Rhythm of Words Clara delves deeper into the rhythmic patterns of speech and poetry, using Fourier analysis to uncover the hidden structures that make stories resonate with audiences.

Episode 3: The Stress Test Clara applies her mathematical model to analyze the stress patterns in famous speeches, revealing the secrets behind their emotional impact and effectiveness.

Episode 4: The Melody of Speech Clara explores the intonation contours of different languages and dialects, using quadratic functions to model the pitch variations that convey meaning and emotion.

Episode 5: The Prosody Revolution Clara's Prosody Analysis Equation gains widespread recognition, inspiring a new generation of artists to embrace mathematical principles in their storytelling. Numeria becomes a hub of artistic excellence, with Clara at the forefront of the prosody revolution.

Conclusion

Math and prosody in storytelling are deeply interconnected, with mathematical principles providing a framework for understanding X.

CHAPTER 17

A microfiction story with less than 299 words that highlights the relationship of math to an epic poem about practical ways and means to live a good interesting productive sustainable and happy life that benefits a great nation, then apply a compactify function to reveal quintessential.

***

Microfiction Story (299 words)

In the heart of a great nation, there lived a mathematician named Elara. She believed that the principles of math could guide people to live good, interesting, productive, sustainable, and happy lives. Inspired by an ancient epic poem, she set out to create a mathematical framework that embodied these ideals.

Elara's journey began with the concept of balance. She saw that just as equations must balance, so must life. She devised a formula: $B=WL$, where B represented balance, W was work, and L was leisure. By maintaining this balance, one could achieve a harmonious life.

Next, Elara focused on growth. She introduced the concept of exponential growth, $G=e^{rt}$, where G was growth, e was the base of natural logarithms, r was the rate of growth, and t was time. She encouraged people to invest in their personal and professional development, knowing that small, consistent efforts would lead to significant progress over time.

Elara also emphasized sustainability. She created a model for sustainable living: $S=RC$, where S was sustainability, R was resources, and C was consumption. By ensuring that consumption never exceeded resources, one could live in harmony with the environment.

Finally, Elara highlighted the importance of happiness. She proposed a happiness equation: $H=PE$, where H was happiness, P was positive experiences, and E was expectations. By managing expectations and seeking positive experiences, one could achieve lasting happiness.

Elara's mathematical framework became a guiding light for the nation, helping its people lead fulfilling lives. Her work proved that math and poetry, though seemingly different, could come together to create a better world.

Compact to 250 words

In a great nation, mathematician Elara believed math could guide people to live good, interesting, productive, sustainable, and happy lives. Inspired by an ancient epic poem, she created a mathematical framework embodying these ideals.

Elara began with balance: B=WL (balance = work/leisure). Maintaining this balance led to a harmonious life.

Next, she focused on growth: G=ert (growth = exponential growth). Investing in personal and professional development led to significant progress over time.

Elara emphasized sustainability: S=RC (sustainability = resources/consumption). Ensuring consumption never exceeded resources led to harmony with the environment.

Finally, she highlighted happiness: H=PE (happiness = positive experiences/expectations). Managing expectations and seeking positive experiences led to lasting happiness.

Elara's framework became a guiding light for the nation, proving that math and poetry could create a better world.

Compact to 200 words

In a great nation, mathematician Elara believed math could guide people to live good, interesting, productive, sustainable, and happy lives. Inspired by an ancient epic poem, she created a mathematical framework embodying these ideals.

Elara began with balance: B=WL (balance = work/leisure). Maintaining this balance led to a harmonious life.

Next, she focused on growth: G=ert (growth = exponential growth). Investing in personal and professional development led to significant progress over time.

Elara emphasized sustainability: S=RC (sustainability = resources/consumption). Ensuring consumption never exceeded resources led to harmony with the environment.

Finally, she highlighted happiness: H=PE (happiness = positive experiences/expectations). Managing expectations and seeking positive experiences led to lasting happiness.

Elara's framework became a guiding light for the nation, proving that math and poetry could create a better world.

Compact to 150 words

In a great nation, mathematician Elara believed math could guide people to live good, interesting, productive, sustainable, and happy lives. Inspired by an X

now compact to under 100 words, then compact to 75 words, then compact to 50, then compact to 25 words, then

20, then 15, then 10, then 5, then 4, then 3, then 2, then 1, then 0 and show as a series of math functions.

CHAPTER 18

Now list symbols used in both math and language, then sort symbols by purpose or function, then create new math symbols, explain their function and benefits of using them, give plenty of examples of using them.

And let's break this down step by step.

Symbols Used in Both Math and Language

Common Symbols

1. **Plus (+)**: Addition in math, conjunction in language.

2. **Minus (-)**: Subtraction in math, hyphen in language.

3. **Equals (=)**: Equality in math, equivalence in language.

4. *Asterisk ()**: Multiplication in math, footnote or emphasis in language.

5. **Slash (/)**: Division in math, separation or alternatives in language.

6. **Parentheses ( )**: Grouping in math, additional information in language.

7. **Colon (:)**: Ratio in math, introduction or explanation in language.

8. **Comma (,)**: Decimal point in some countries in math, separation of items in language.

9. **Period (.)**: Decimal point in math, end of a sentence in language.

Sorting Symbols by Purpose or Function

Arithmetic Operations

- **Plus (+)**
- **Minus (-)**
- *Asterisk ()**
- **Slash (/)**
- **Equals (=)**

Grouping and Separation

- **Parentheses ( )**
- **Comma (,)**
- **Period (.)**

Additional Information

- *Asterisk ()**
- **Colon (:)**

Creating New Math Symbols

Symbol: ⊕ **(Circle Plus)**

Function: Represents a special type of addition where the result is always positive. Benefits: Simplifies expressions involving absolute values and ensures non-negative results.

Examples:

1. 3⊕5=8
2. −4⊕7=3
3. −6⊕−2=8

Symbol: ⊗ **(Circle Times)**

Function: Represents a special type of multiplication where the result is always positive. Benefits: Simplifies expressions involving absolute values and ensures non-negative results.

Examples:

1. 3⊗5=15
2. −4⊗7=28
3. −6⊗−2=12

Symbol: ⊖ **(Circle Minus)**

Function: Represents a special type of subtraction where the result is always positive. Benefits: Simplifies expressions involving absolute values and ensures non-negative results.

Examples:

1. 8⊖3=5
2. 7⊖−4=11
3. −2⊖−6=4

Benefits of Using New Symbols

1. **Clarity**: These symbols make it clear that the operations result in non-negative values.
2. **Simplicity**: They simplify expressions involving absolute values, reducing the need for additional notation.
3. **Consistency**: They provide a consistent way to handle operations that require non-negative results.

By introducing these new symbols, we can make mathematical expressions more intuitive and easier to work with, especially in fields where non-negative values are crucial, such as probability and statistics.

And you can create more new math symbols?

Yes.

Let's go.

New Math Symbols

Symbol: ⊛ (Star Product)

Function: Represents a special type of multiplication where the result is the product of the absolute values of the operands. Benefits: Simplifies expressions involving absolute values and ensures non-negative results.

Examples:

1. 3⊛5=15
2. −4⊛7=28
3. −6⊛−2=12

Symbol: ⊖ (Circle Division)

Function: Represents a special type of division where the result is always positive. Benefits: Simplifies expressions involving absolute values and ensures non-negative results.

Examples:

1. 15⊖3=5
2. −28⊖7=4
3. −12⊖−2=6

Symbol: ⊖ (Circle Exponentiation)

Function: Represents a special type of exponentiation where the result is always positive. Benefits: Simplifies expressions involving absolute values and ensures non-negative results.

Examples:

1. $3 ⊖ 2 = 9$
2. $-4 ⊖ 2 = 16$
3. $-2 ⊖ 3 = 8$

Benefits of Using New Symbols

1. **Clarity**: These symbols make it clear that the operations result in non-negative values.

2. **Simplicity**: They simplify expressions involving absolute values, reducing the need for additional notation.

3. **Consistency**: They provide a consistent way to handle operations that require non-negative results.

By introducing these new symbols, we can make mathematical expressions more intuitive and easier to work with, especially in fields where non-negative values are crucial, such as probability and statistics.

What do you think?

Ready to start a new Renaissance Golden Age of Math as well as the storytelling system. and here are a few examples that showcase its elegance and beauty:

1. The Fibonacci Sequence

The Fibonacci sequence is a series of numbers where each number is the sum of the two preceding ones, starting from 0 and 1. It goes like this: 0, 1, 1, 2, 3, 5, 8, 13, 21, and so on. This sequence appears in nature, such as in the arrangement of leaves on a stem, the branching of trees, and the spirals of shells. The poetic beauty lies in its simplicity and its connection to the natural world.

2. Euler's Identity

Euler's identity is often considered one of the most beautiful equations in mathematics: $$e^{i\pi} + 1 = 0$$
This equation connects five of the most important numbers in mathematics: e (the base of natural logarithms), i (the imaginary unit), π (pi, the ratio of a circle's circumference to its diameter), 1, and 0. The elegance of this equation lies in its simplicity and the profound relationships it reveals between these fundamental constants.

3. The Golden Ratio

The golden ratio, often denoted by the Greek letter φ (phi), is approximately equal to 1.6180339887. It can be found by dividing a line into two parts such that the ratio of the whole line to the longer part is the same as the ratio of the longer part to the shorter part. The golden ratio appears in art, architecture, and nature, and is often associated with aesthetic beauty. Its poetic nature comes from its ubiquity and its pleasing proportions.

4. The Mandelbrot Set

The Mandelbrot set is a set of complex numbers that produces a fractal when plotted. The fractal is infinitely complex and exhibits self-similarity, meaning that smaller parts of the fractal resemble the whole. The beauty of the Mandelbrot set lies in its intricate patterns and the way it reveals the complexity and order within chaos.

5. Pythagorean Theorem

The Pythagorean theorem states that in a right-angled triangle, the square of the length of the hypotenuse (the side opposite the right angle) is equal to the sum of the squares of the lengths of the other two sides: $$a^2 + b^2 = c^2$$ This theorem is not only fundamental to geometry but also has a certain poetic simplicity. It elegantly connects the sides of a right-angled triangle and has numerous applications in various fields.

What Makes Math Poetic?

- **Simplicity and Elegance**: Many mathematical concepts are simple yet profound, revealing deep truths with minimal complexity.

- **Universality**: Math transcends cultural and linguistic barriers, making it a universal language that everyone can understand.

- **Connection to Nature**: Many mathematical patterns and relationships are found in the natural world, highlighting the inherent order and beauty of the universe.

- **Aesthetic Appeal**: Certain mathematical concepts, like the golden ratio and fractals, have an inherent aesthetic appeal that resonates with our sense of beauty.

And math, in its purest form, is a way of understanding and describing the world around us. Its poetic nature lies in its ability to reveal the hidden patterns and structures that underpin reality.

Yes.

Yes.

Yes.

I agree, as math truly has a unique way of revealing the beauty and order in the world around us as a hidden language that, once understood, opens up a whole new perspective on reality.

Now can we have some fun with math and poem?

***

Absolutely!

And let's blend the precision of math with the creativity of a mathematical poem for you:

The Dance of Numbers

In the realm where numbers dance, Equations form a sweet romance. With every digit, line, and curve, Mathematics finds its nerve.

Zero stands, a silent knight, Guarding secrets in the night. One steps forward, bold and true, Leading others in the queue.

Two and three, they intertwine, Creating patterns so divine. Four and five, they take their place, In this endless, cosmic race.

Six and seven, partners strong, In this dance, they can't go wrong. Eight and nine, they twirl and spin, In the world where math begins.

Pi, the circle's hidden key, Unlocks the mysteries we see. Euler's identity, so neat, Brings together math's heartbeat.

And in this dance of numbers bright, Math and poetry unite. For in their union, we find grace, A timeless beauty, we embrace.

The relationship between math and poetry: Less than 75 Words, Boil down to the core message. Example: "I wanted to finish a marathon to prove to my kids that anyone can do it. Despite injuring my wrist a week before, I completed the marathon, showing that perseverance pays off." Less than 25 Words Capture the essence in a single sentence. Example: "Injured a week before, I finished the marathon to prove to my kids that perseverance pays off." Less than 19 Words Further distill the message. Example: "Injured before the marathon, I finished to show my kids perseverance pays off."6 Words Example: "Injured, finished marathon, proved perseverance."5 Words Example: "Injured, finished marathon, proved perseverance."4 Words Example: "Injured, finished, proved perseverance."3 Words Example: "Finished, proved perseverance."2 Words Example: "Proved perseverance."1 Word Example: "Perseverance."0 Words Example: Silence can sometimes be the most powerful statement. Math Functions Representation1.Clarity: ($f(x) = \text{clear}(x)$)2.Dynamism: ($g(x) = \text{dynamic}(x)$)3.Efficiency: ($h(x) = \text{efficient}(x)$)4.Effectiveness: ($i(x) = \text{effective}(x)$)5.Realism: ($j(x) = \text{realistic}(x)$)Combining these functions: [$S(x) = f(x) + g(x) + h(x) + i(x) + j(x)$]

This function ($S(x)$) represents an exceptional story that embodies clarity, dynamism, efficiency, effectiveness, and realism?

And is it an example beautifully illustrates the relationship between math and poetry by showing how both can distill complex ideas into their essence?

Now state:

Less than 75 Words

Boil down to the core message. Example: "I wanted to finish a marathon to prove to my kids that anyone can do it. Despite injuring my wrist a week before, I completed the marathon, showing that perseverance pays off."

Less than 25 Words

Capture the essence in a single sentence. Example: "Injured a week before, I finished the marathon to prove to my kids that perseverance pays off."

Less than 19 Words

Further distill the message. Example: "Injured before the marathon, I finished to show my kids perseverance pays off."

6 Words

Example: "Injured, finished marathon, proved perseverance."

5 Words

Example: "Injured, finished marathon, proved perseverance."

4 Words

Example: "Injured, finished, proved perseverance."

3 Words

Example: "Finished, proved perseverance."

2 Words

Example: "Proved perseverance."

1 Word

Example: "Perseverance."

0 Words

Example: Silence can sometimes be the most powerful statement.

Math Functions Representation

1. **Clarity**: f(x)=clear(x)
2. **Dynamism**: g(x)=dynamic(x)
3. **Efficiency**: h(x)=efficient(x)
4. **Effectiveness**: i(x)=effective(x)
5. **Realism**: j(x)=realistic(x)

Combining these functions: \[S(x) = f(x) + g(x) + h(x) + i(x) + j(x) \]

This function S(x) represents an exceptional story that embodies clarity, dynamism, efficiency, effectiveness, and realism.

And does this example show how both math and poetry can distill complex ideas into their essence, capturing the core message in a clear, dynamic, efficient, effective, and realistic way?

Next example:

The YouTube page features the full movie "Tank Force," a WWII film about British soldiers escaping a prisoner-of-war camp in Libya. The movie, directed by Terence Young, showcases the harsh desert conditions and the soldiers' struggle for survival. The comments section includes various insights, such as the director's background as a tank commander and connections to the James Bond film series. Viewers share their thoughts on the film's historical inaccuracies, production quality, and overall entertainment value. Some comments also highlight the film's portrayal of relationships between British, German, Italian, and Arab characters during the war.

Next example:

the relationship of math to an epic poem about practical ways and means to live a good, interesting, productive, sustainable, and happy life that benefits a great nation.

Then, we'll compact it step by step and express it as a series of poetic math functions.

Microfiction Story (299 words)

In the heart of a great nation, there lived a mathematician named Ada. She believed that the principles of math could guide people to live fulfilling lives. Ada wrote an epic poem, "The Equation of Life," which outlined practical ways to achieve happiness and sustainability.

Her poem began with the basics: "Add kindness, subtract greed, multiply love, and divide resources equally." She used mathematical metaphors to illustrate the importance of balance and harmony. "Find the common denominator in conflicts, and solve for peace," she wrote.

Ada's poem became a guiding light for the nation. People started to see the beauty in numbers and the wisdom in equations. Farmers used math to optimize their crops, ensuring sustainable practices. Teachers incorporated math into their lessons on ethics and community. Engineers designed efficient systems that benefited everyone.

The nation thrived, and Ada's poem was celebrated as a masterpiece. It was proof that math and poetry could work together to create a better world.

Compacting the Story

250 Words

Ada, a mathematician, believed math could guide people to fulfilling lives. She wrote "The Equation of Life," an epic poem with practical ways to achieve happiness and sustainability. Using mathematical metaphors, she illustrated balance and harmony: "Add kindness, subtract greed, multiply love, and divide resources equally." Her poem became a guiding light, inspiring farmers, teachers, and engineers to use math for the greater good. The nation thrived, proving that math and poetry could create a better world.

200 Words

Mathematician Ada believed math could guide people to fulfilling lives. She wrote "The Equation of Life," an epic poem with practical ways to achieve happiness and sustainability. Using mathematical metaphors, she illustrated balance and harmony: "Add kindness, subtract greed, multiply love, and divide resources equally." Her poem inspired farmers, teachers, and engineers to use math for the greater good. The nation thrived, proving that math and poetry could create a better world.

150 Words

Ada, a mathematician, wrote "The Equation of Life," an epic poem with practical ways to achieve happiness and sustainability. Using mathematical metaphors, she illustrated balance and harmony: "Add kindness, subtract greed, multiply love, and divide resources equally." Her poem inspired farmers, teachers, and engineers to use math for the greater good. The nation thrived, proving that math and poetry could create a better world.

100 Words

Mathematician Ada wrote "The Equation of Life," an epic poem with practical ways to achieve happiness and sustainability. Using mathematical metaphors, she illustrated balance and harmony: "Add kindness, subtract greed, multiply love, and divide resources equally." Her poem inspired farmers, teachers, and engineers to use math for the greater good. The nation thrived, proving that math and poetry could create a better world.

50 Words

Ada's poem, "The Equation of Life," used math metaphors to guide people to happiness and sustainability: "Add kindness, subtract greed, multiply love, and divide resources equally." It inspired farmers, teachers, and engineers. The nation thrived, proving math and poetry could create a better world.

25 Words

Ada's poem, "The Equation of Life," used math metaphors to guide people to happiness and sustainability. It inspired many, proving math and poetry could create a better world.

20 Words

Ada's poem, "The Equation of Life," used math metaphors to guide people. It inspired many, proving math and poetry's power.

15 Words

Ada's poem used math metaphors to guide people, proving math and poetry's power to inspire.

10 Words

Ada's poem used math to inspire, proving its power.

7 Words

Ada's poem: math inspires, proves its power.

5 Words

Math inspires, proves its power.

4 Words

Math inspires, proves power.

3 Words

Math inspires power.

2 Words

Math inspires.

1 Word

Inspiration.

0 Words

(Silence)

Explanation

The story illustrates how math can be used as a metaphorical language to convey practical ways to live a good, interesting, productive, sustainable, and happy life. By compacting the story, we distill its essence, showing how the core message remains powerful even in its simplest form.

Detailed Comments

- **Clarity**: The story and its compacted versions maintain a clear message about the power of math and poetry.

- **Dynamism**: The use of mathematical metaphors adds a dynamic element to the narrative.

- **Efficiency**: The compacted versions efficiently convey the core message without losing its essence.

- **Effectiveness**: Each version effectively communicates the relationship between math and poetry.

- **Realism**: The story is grounded in realistic applications of math in everyday life.

Series of Opinions

1. Math and poetry can complement each other beautifully.
2. Mathematical metaphors can simplify complex ideas.
3. Compacting stories can reveal their core essence.
4. Math can guide practical ways to live a fulfilling life.
5. Poetry can make mathematical concepts more relatable.

Poetic Math Functions

1. **Clarity**: $f(x)=clear(x)$
2. **Dynamism**: $g(x)=dynamic(x)$
3. **Efficiency**: $h(x)=efficient(x)$
4. **Effectiveness**: $i(x)=effective(x)$
5. **Realism**: $j(x)=realistic(x)$

Combining these functions: $$S(x) = f(x) + g(x) + h(x) + i(x) + j(x)$$

This function S(x) represents an exceptional story that embodies clarity, dynamism, efficiency, effectiveness, and realism?

***

Next example:

Microfiction Story (299 words)

In the heart of a thriving nation, there lived a visionary leader named Elara. She believed that the principles of mathematics could guide the nation to greatness. One day, she stood before her people to deliver an epic speech titled "The Geometry of Greatness."

"To build a truly great nation," Elara began, "we must understand the fundamental principles of math. Just as a triangle is the strongest shape, our nation must be built on three pillars: equality, innovation, and sustainability."

She continued, "Equality is like the base of our triangle. It provides stability and support. We must ensure that every citizen has equal opportunities and access to resources. Innovation is the height of our triangle, reaching for the stars. We must encourage creativity and technological advancements to propel us forward. Sustainability is the third side, connecting equality and innovation. We must use our resources wisely to ensure a prosperous future for generations to come."

Elara's words resonated deeply with the people. They began to see the beauty in numbers and the wisdom in equations. Farmers used math to optimize their crops, ensuring sustainable practices. Teachers incorporated math into their lessons on ethics and community. Engineers designed efficient systems that benefited everyone.

The nation thrived, and Elara's speech became a guiding light. It was proof that math and visionary leadership could work together to create a better world.

Compacting the Story

250 Words

In a thriving nation, visionary leader Elara believed math could guide the nation to greatness. In her epic speech, "The Geometry of Greatness," she explained that a great nation must be built on three pillars: equality, innovation, and sustainability.

"Equality is like the base of our triangle," Elara said. "It provides stability and support. We must ensure that every citizen has equal opportunities and access to resources. Innovation is the height of our triangle, reaching for the stars. We must encourage creativity and technological advancements. Sustainability is the third side, connecting equality and innovation. We must use our resources wisely to ensure a prosperous future."

Elara's words resonated deeply. Farmers optimized their crops, teachers incorporated math into lessons, and engineers designed efficient systems. The nation thrived, proving that math and visionary leadership could create a better world.

200 Words

Visionary leader Elara believed math could guide her nation to greatness. In her speech, "The Geometry of Greatness," she explained that a great nation must be built on three pillars: equality, innovation, and sustainability.

"Equality is like the base of our triangle," Elara said. "It provides stability. We must ensure equal opportunities for all. Innovation is the height, reaching for the stars. We must encourage creativity and technological advancements. Sustainability is the third side, connecting equality and innovation. We must use resources wisely for a prosperous future."

Elara's words resonated deeply. Farmers optimized crops, teachers incorporated math into lessons, and engineers designed efficient systems. The nation thrived, proving that math and visionary leadership could create a better world.

150 Words

Elara, a visionary leader, believed math could guide her nation to greatness. In her speech, "The Geometry of Greatness," she explained that a great nation must be built on three pillars: equality, innovation, and sustainability.

"Equality is the base of our triangle," Elara said. "It provides stability. We must ensure equal opportunities for all. Innovation is the height, reaching for the stars. We must encourage creativity and technological advancements. Sustainability is the third side, connecting equality and innovation. We must use resources wisely for a prosperous future."

Elara's words resonated deeply. Farmers optimized crops, teachers incorporated math into lessons, and engineers designed efficient systems. The nation thrived, proving that math and visionary leadership could create a better world.

100 Words

Elara, a visionary leader, believed math could guide her nation to greatness. In her speech, "The Geometry of

Greatness," she explained that a great nation must be built on three pillars: equality, innovation, and sustainability.

"Equality is the base of our triangle," Elara said. "It provides stability. We must ensure equal opportunities for all. Innovation is the height, reaching for the stars. We must encourage creativity and technological advancements. Sustainability is the third side, connecting equality and innovation. We must use resources wisely for a prosperous future."

Elara's words resonated deeply. The nation thrived, proving that math and visionary leadership could create a better world.

50 Words

Elara's speech, "The Geometry of Greatness," explained that a great nation must be built on three pillars: equality, innovation, and sustainability. Equality provides stability, innovation reaches for the stars, and sustainability connects them. Her words resonated deeply, proving that math and visionary leadership could create a better world.

25 Words

Elara's speech, "The Geometry of Greatness," explained that equality, innovation, and sustainability are key. Her words resonated, proving math and visionary leadership could create a better world.

20 Words

Elara's speech explained that equality, innovation, and sustainability are key. Her words proved math and visionary leadership could create a better world.

15 Words

Elara's speech: equality, innovation, and sustainability are key. Math and visionary leadership create a better world.

10 Words

Elara's speech: equality, innovation, sustainability. Math and leadership create a better world.

7 Words

Equality, innovation, sustainability. Math creates a better world.

5 Words

Math creates a better world.

4 Words

Math improves the world.

3 Words

Math inspires greatness.

2 Words

Math inspires.

1 Word

Inspiration.

0 Words

(Silence)

Explanation

The story illustrates how math can be used as a metaphorical language to convey practical ways to create a great nation. By compacting the story, we distill its essence, showing how the core message remains powerful even in its simplest form.

Detailed Comments

- **Clarity**: The story and its compacted versions maintain a clear message about the power of math and visionary leadership.
- **Dynamism**: The use of mathematical metaphors adds a dynamic element to the narrative.
- **Efficiency**: The compacted versions efficiently convey the core message without losing its essence.

- **Effectiveness**: Each version effectively communicates the relationship between math and leadership.
- **Realism**: The story is grounded in realistic applications of math in nation-building.

Series of Opinions

1. Math and leadership can complement each other beautifully.
2. Mathematical metaphors can simplify complex ideas.
3. Compacting stories can reveal their core essence.
4. Math can guide practical ways to build a great nation.
5. Visionary leadership can make mathematical concepts more relatable.

Poetic Math Functions

1. **Clarity**: $f(x)=clear(x)$
2. **Dynamism**: $g(x)=dynamic(x)$
3. **Efficiency**: $h(x)=efficient(x)$
4. **Effectiveness**: $i(x)=effective(x)$

5. **Realism**: j(x)=realistic(x)

Combining these functions: $$ S(x) = f(x) + g(x) + h(x) + i(x) + j(x) $$

This function S(x) represents an exceptional story that embodies clarity, dynamism, efficiency, effectiveness, and realism.

***

Next example:

Microfiction Story (299 words)

In the bustling city of Numeria, lived a mathematician named Leo. He was known for his groundbreaking work in algorithms and equations. But Leo was exhausted. His mind was a constant whirl of numbers and formulas, and he longed for a break.

One evening, Leo stood before the city council to make a plea. "Mathematics teaches us the importance of balance," he began. "Just as an equation must be balanced, so must our lives. We cannot continue to work endlessly without rest. We need a vacation, a real one, to rejuvenate our minds and bodies."

Leo continued, "Consider the Fibonacci sequence. Each number is the sum of the two preceding ones, creating a harmonious pattern. Our lives should follow a similar pattern of work and rest. Without rest, we disrupt this harmony and risk burnout."

The council members, intrigued by Leo's mathematical analogy, began to see the wisdom in his words. They realized that just as numbers need balance, so do people. They approved a city-wide vacation, encouraging everyone to take time off to relax and recharge.

Leo spent his vacation in the serene mountains, away from the noise of the city. He hiked, read books, and let his mind wander freely. When he returned, he was refreshed and full of new ideas. The city of Numeria thrived, proving that balance, as taught by mathematics, was essential for a happy and productive life.

Compacting the Story

250 Words

In Numeria, mathematician Leo was known for his groundbreaking work. But Leo was exhausted and longed for a break. One evening, he stood before the city council to make a plea. "Mathematics teaches us the importance of balance," he began. "Just as an equation must be balanced, so must our lives. We need a vacation to rejuvenate our minds and bodies."

Leo continued, "Consider the Fibonacci sequence. Each number is the sum of the two preceding ones, creating a harmonious pattern. Our lives should follow a similar pattern of work and rest. Without rest, we disrupt this harmony and risk burnout."

The council members, intrigued by Leo's mathematical analogy, saw the wisdom in his words. They approved a city-wide vacation, encouraging everyone to take time off to relax and recharge.

Leo spent his vacation in the serene mountains, away from the city's noise. He hiked, read books, and let his mind wander freely. When he returned, he was refreshed and full of new ideas. Numeria thrived, proving that balance, as taught by mathematics, was essential for a happy and productive life.

200 Words

In Numeria, mathematician Leo was known for his groundbreaking work. But Leo was exhausted and longed for a break. One evening, he stood before the city council to make a plea. "Mathematics teaches us the importance of balance," he began. "Just as an equation must be balanced, so must our lives. We need a vacation to rejuvenate our minds and bodies."

Leo continued, "Consider the Fibonacci sequence. Each number is the sum of the two preceding ones, creating a harmonious pattern. Our lives should follow a similar pattern of work and rest. Without rest, we disrupt this harmony and risk burnout."

The council members, intrigued by Leo's mathematical analogy, saw the wisdom in his words. They approved a city-wide vacation, encouraging everyone to take time off to relax and recharge.

Leo spent his vacation in the serene mountains, away from the city's noise. He hiked, read books, and let his mind wander freely. When he returned, he was refreshed and full of new ideas. Numeria thrived, proving that balance, as taught by mathematics, was essential for a happy and productive life.

150 Words

In Numeria, mathematician Leo was known for his groundbreaking work. But Leo was exhausted and longed for a break. He stood before the city council to make a plea. "Mathematics teaches us the importance of balance," he began. "Just as an equation must be balanced, so must our lives. We need a vacation to rejuvenate our minds and bodies."

Leo continued, "Consider the Fibonacci sequence. Each number is the sum of the two preceding ones, creating a harmonious pattern. Our lives should follow a similar pattern of work and rest. Without rest, we disrupt this harmony and risk burnout."

The council members saw the wisdom in Leo's words. They approved a city-wide vacation, encouraging everyone to take time off to relax and recharge.

Leo spent his vacation in the serene mountains. When he returned, he was refreshed and full of new ideas. Numeria thrived, proving that balance, as taught by mathematics, was essential for a happy and productive life.

100 Words

In Numeria, mathematician Leo was known for his groundbreaking work. But Leo was exhausted and longed for a break. He stood before the city council to make a plea. "Mathematics teaches us the importance of balance," he began. "Just as an equation must be balanced, so must our lives. We need a vacation to rejuvenate our minds and bodies."

The council members saw the wisdom in Leo's words. They approved a city-wide vacation, encouraging everyone to take time off to relax and recharge.

Leo spent his vacation in the mountains. When he returned, he was refreshed and full of new ideas. Numeria thrived.

50 Words

In Numeria, mathematician Leo was exhausted. He pleaded for a vacation, explaining that balance, like in math, was essential. The council approved a city-wide break. Leo spent his vacation in the mountains. Refreshed, he returned with new ideas. Numeria thrived, proving balance was key to productivity.

25 Words

Mathematician Leo pleaded for a vacation, explaining balance was essential. The council approved. Refreshed, Leo returned with new ideas. Numeria thrived, proving balance was key.

20 Words

Leo pleaded for a vacation, explaining balance was essential. The council approved. Refreshed, Leo returned with new ideas. Numeria thrived.

15 Words

Leo pleaded for a vacation. The council approved. Refreshed, Leo returned with new ideas. Numeria thrived.

10 Words

Leo pleaded for vacation. Refreshed, he returned. Numeria thrived.

7 Words

Leo pleaded. Refreshed, he returned. Numeria thrived.

5 Words

Leo pleaded. Numeria thrived.

4 Words

Leo pleaded. Thrived.

3 Words

Leo pleaded.

2 Words

Leo pleaded.

1 Word

Pleaded.

0 Words

(Silence)

Explanation

The story illustrates how math can be used as a metaphorical language to convey the importance of balance and rest. By compacting the story, we distill its essence, showing how the core message remains powerful even in its simplest form.

Detailed Comments

- **Clarity**: The story and its compacted versions maintain a clear message about the importance of balance and rest.
- **Dynamism**: The use of mathematical metaphors adds a dynamic element to the narrative.
- **Efficiency**: The compacted versions efficiently convey the core message without losing its essence.
- **Effectiveness**: Each version effectively communicates the relationship between math and the need for rest.
- **Realism**: The story is grounded in realistic applications of math in everyday life.

Series of Opinions

1. Math and rest can complement each other beautifully.
2. Mathematical metaphors can simplify complex ideas.
3. Compacting stories can reveal their core essence.

4. Math can guide practical ways to achieve balance and rest.
5. Rest is essential for productivity and creativity.

Poetic Math Functions

1. **Clarity**: $f(x)=\text{clear}(x)$
2. **Dynamism**: $g(x)=\text{dynamic}(x)$
3. **Efficiency**: $h(x)=\text{efficient}(x)$
4. **Effectiveness**: $i(x)=\text{effective}(x)$
5. **Realism**: $j(x)=\text{realistic}(x)$

Combining these functions: $$S(x) = f(x) + g(x) + h(x) + i(x) + j(x)$$

This function $S(x)$ represents an exceptional story that embodies clarity, dynamism, efficiency, effectiveness, and realism.

***

And here is another microfiction story (299 words):

In the city of Numeria, a brilliant mathematician named Iris discovered a unique formula to identify greedy and dangerous people. She called it the "Greed Quotient" (GQ). The formula was simple: GQ = (Desire for Wealth + Willingness to Harm Others) / (Empathy + Integrity).

Iris presented her findings to the city council, explaining how the Greed Quotient could help protect the citizens. "By using this formula," she said, "we can identify those who pose a threat to our community and take preventive measures."

The council was skeptical but agreed to a trial. They applied the formula to various individuals, and the results were astonishing. Those with high GQ scores were indeed involved in unethical activities, while those with low scores were upstanding citizens.

The city implemented the Greed Quotient as a standard measure, and soon, Numeria became a safer place. People began to value empathy and integrity more, knowing that their actions were being measured.

One day, a wealthy businessman named Victor, known for his ruthless tactics, was found to have a high GQ score. The council confronted him, and he was forced to change his ways or face severe consequences. Victor chose to reform, and his transformation inspired others to follow suit.

Iris's formula not only protected Numeria but also fostered a culture of kindness and integrity. The city thrived, proving that math could be a powerful tool in creating a just and harmonious society.

Compacting the Story

250 Words

In Numeria, mathematician Iris discovered a formula to identify greedy and dangerous people, called the "Greed Quotient" (GQ): GQ = (Desire for Wealth + Willingness to Harm Others) / (Empathy + Integrity).

Iris presented her findings to the city council, explaining how the Greed Quotient could protect citizens. "By using this formula," she said, "we can identify those who pose a threat to our community and take preventive measures."

The council agreed to a trial. They applied the formula to various individuals, and the results were astonishing. Those with high GQ scores were involved in unethical activities, while those with low scores were upstanding citizens.

Numeria implemented the Greed Quotient, and soon, the city became safer. People valued empathy and integrity more, knowing their actions were measured.

A wealthy businessman named Victor, known for his ruthless tactics, had a high GQ score. The council confronted him, and he was forced to change his ways or face severe consequences. Victor chose to reform, inspiring others to follow suit.

Iris's formula protected Numeria and fostered a culture of kindness and integrity. The city thrived, proving that math could be a powerful tool in creating a just and harmonious society.

200 Words

In Numeria, mathematician Iris discovered a formula to identify greedy and dangerous people, called the "Greed Quotient" (GQ): GQ = (Desire for Wealth + Willingness to Harm Others) / (Empathy + Integrity).

Iris presented her findings to the city council, explaining how the Greed Quotient could protect citizens. The council agreed to a trial. They applied the formula to various individuals, and the results were astonishing. Those with high GQ scores were involved in unethical activities, while those with low scores were upstanding citizens.

Numeria implemented the Greed Quotient, and soon, the city became safer. People valued empathy and integrity more, knowing their actions were measured.

A wealthy businessman named Victor, known for his ruthless tactics, had a high GQ score. The council confronted him, and he was forced to change his ways or face severe consequences. Victor chose to reform, inspiring others to follow suit.

Iris's formula protected Numeria and fostered a culture of kindness and integrity. The city thrived, proving that math could be a powerful tool in creating a just and harmonious society.

150 Words

In Numeria, mathematician Iris discovered a formula to identify greedy and dangerous people, called the "Greed Quotient" (GQ): GQ = (Desire for Wealth + Willingness to Harm Others) / (Empathy + Integrity).

Iris presented her findings to the city council, explaining how the Greed Quotient could protect citizens. The council agreed to a trial. They applied the formula to various individuals, and the results were astonishing. Those with high GQ scores were involved in unethical activities, while those with low scores were upstanding citizens.

Numeria implemented the Greed Quotient, and soon, the city became safer. People valued empathy and integrity more, knowing their actions were measured.

A wealthy businessman named Victor, known for his ruthless tactics, had a high GQ score. The council confronted him, and he was forced to change his ways or face severe consequences. Victor chose to reform, inspiring others to follow suit.

Iris's formula protected Numeria and fostered a culture of kindness and integrity.

100 Words

In Numeria, mathematician Iris discovered a formula to identify greedy and dangerous people, called the "Greed Quotient" (GQ): GQ = (Desire for Wealth + Willingness to Harm Others) / (Empathy + Integrity).

Iris presented her findings to the city council, and they agreed to a trial. The results were astonishing. Those with high GQ scores were involved in unethical activities, while those with low scores were upstanding citizens.

Numeria implemented the Greed Quotient, and the city became safer. A ruthless businessman named Victor had a high GQ score. Confronted by the council, he reformed, inspiring others. Numeria thrived.

50 Words

In Numeria, mathematician Iris discovered the "Greed Quotient" (GQ) to identify dangerous people: GQ = (Desire for Wealth + Willingness to Harm Others) / (Empathy + Integrity). The council used it, making the city safer. A ruthless businessman reformed, inspired by the GQ. Numeria thrived.

25 Words

Iris's "Greed Quotient" identified dangerous people. The council used it, making Numeria safer. A ruthless businessman reformed, inspired by the GQ. Numeria thrived.

20 Words

Iris's "Greed Quotient" identified dangerous people. The council used it. A ruthless businessman reformed, inspired by the GQ. Numeria thrived.

15 Words

Iris's "Greed Quotient" identified dangerous people. The council used it. Numeria thrived.

10 Words

Iris's "Greed Quotient" identified danger. Numeria thrived.

7 Words

Iris's formula identified danger. Numeria thrived.

5 Words

Iris's formula saved Numeria.

4 Words

Iris's formula saved.

3 Words

Formula saved.

2 Words

Formula saved.

1 Word

Saved.

0 Words

(Silence)

Explanation

The story illustrates how math can be used as a tool to identify and avoid greedy and dangerous people. By compacting the story, we distill its essence, showing how the core message remains powerful even in its simplest form.

Detailed Comments

- **Clarity**: The story and its compacted versions maintain a clear message about the power of math to identify and avoid danger.

- **Dynamism**: The use of a mathematical formula adds a dynamic element to the narrative.

- **Efficiency**: The compacted versions efficiently convey the core message without losing its essence.
- **Effectiveness**: Each version effectively communicates the relationship between math and safety.
- **Realism**: The story is grounded in realistic applications of math in everyday life.

Series of Opinions

1. Math can be a powerful tool for identifying and avoiding danger.
2. Mathematical formulas can simplify complex ideas.
3. Compacting stories can reveal their core essence.
4. Math can guide practical ways to ensure safety and integrity.
5. Formulas can inspire positive change.

Poetic Math Functions

1. **Clarity**: $f(x)=clear(x)$
2. **Dynamism**: $g(x)=dynamic(x)$
3. **Efficiency**: $h(x)=efficient(x)$
4. **Effectiveness**: $i(x)=effective(x)$

5. **Realism**: $j(x) = realistic(x)$

Combining these functions: $$S(x) = f(x) + g(x) + h(x) + i(x) + j(x)$$

This function $S(x)$ represents an exceptional story that embodies clarity, dynamism, efficiency, effectiveness, and realism.

CHAPTER 19

Microfiction Story (299 words)

In the bustling city of Literaria, a mathematician named Alex dreamed of writing a bestseller novel. He believed that the principles of math could guide him in crafting a compelling story. Alex began by analyzing the structure of successful novels, identifying patterns and common elements.

He discovered that most bestsellers followed a specific formula: a strong opening, rising tension, a climax, and a satisfying resolution. Using this formula, Alex plotted his novel, ensuring each chapter adhered to the mathematical principles he had identified.

Alex also applied statistical analysis to his writing. He studied word frequency and sentence length, aiming for a balance that would keep readers engaged. He used algorithms to predict the emotional impact of different plot twists, ensuring his story would resonate with a wide audience.

As Alex wrote, he meticulously tracked his progress, using graphs and charts to visualize the development of his plot and characters. He adjusted his writing based on feedback, refining his novel until it was a polished masterpiece.

When Alex finally published his book, it became an instant bestseller. Critics praised its perfect pacing, engaging characters, and emotional depth. Readers couldn't put it down, captivated by the seamless blend of math and storytelling.

Alex's success proved that math and creativity were not mutually exclusive. By applying mathematical principles to his writing, he had created a novel that resonated with readers on a profound level. His story became a testament to the power of numbers in crafting a compelling narrative.

Compacting the Story

250 Words

In Literaria, mathematician Alex dreamed of writing a bestseller novel. He believed math could guide him in crafting a compelling story. Alex analyzed successful novels, identifying patterns and common elements.

He discovered that most bestsellers followed a specific formula: a strong opening, rising tension, a climax, and a satisfying resolution. Using this formula, Alex plotted his novel, ensuring each chapter adhered to the mathematical principles he had identified.

Alex also applied statistical analysis to his writing. He studied word frequency and sentence length, aiming for a balance that would keep readers engaged. He used algorithms to predict the emotional impact of different plot twists, ensuring his story would resonate with a wide audience.

As Alex wrote, he meticulously tracked his progress, using graphs and charts to visualize the development of his plot and characters. He adjusted his writing based on feedback, refining his novel until it was a polished masterpiece.

When Alex finally published his book, it became an instant bestseller. Critics praised its perfect pacing, engaging characters, and emotional depth. Readers couldn't put it down, captivated by the seamless blend of math and storytelling.

Alex's success proved that math and creativity were not mutually exclusive. By applying mathematical principles to his writing, he had created a novel that resonated with readers on a profound level.

200 Words

In Literaria, mathematician Alex dreamed of writing a bestseller novel. He believed math could guide him in crafting a compelling story. Alex analyzed successful novels, identifying patterns and common elements.

He discovered that most bestsellers followed a specific formula: a strong opening, rising tension, a climax, and a satisfying resolution. Using this formula, Alex plotted his novel, ensuring each chapter adhered to the mathematical principles he had identified.

Alex also applied statistical analysis to his writing. He studied word frequency and sentence length, aiming for a balance that would keep readers engaged. He used algorithms to predict the emotional impact of different plot twists, ensuring his story would resonate with a wide audience.

As Alex wrote, he meticulously tracked his progress, using graphs and charts to visualize the development of his plot and characters. He adjusted his writing based on feedback, refining his novel until it was a polished masterpiece.

When Alex finally published his book, it became an instant bestseller. Critics praised its perfect pacing, engaging characters, and emotional depth. Readers couldn't put it down, captivated by the seamless blend of math and storytelling.

Alex's success proved that math and creativity were not mutually exclusive.

150 Words

In Literaria, mathematician Alex dreamed of writing a bestseller novel. He believed math could guide him in crafting a compelling story. Alex analyzed successful novels, identifying patterns and common elements.

He discovered that most bestsellers followed a specific formula: a strong opening, rising tension, a climax, and a satisfying resolution. Using this formula, Alex plotted his novel, ensuring each chapter adhered to the mathematical principles he had identified.

Alex also applied statistical analysis to his writing. He studied word frequency and sentence length, aiming for a balance that would keep readers engaged. He used algorithms to predict the emotional impact of different plot twists.

As Alex wrote, he meticulously tracked his progress, using graphs and charts to visualize the development of his plot and characters. He adjusted his writing based on feedback, refining his novel until it was a polished masterpiece.

When Alex published his book, it became an instant bestseller. Critics praised its perfect pacing and emotional depth.

100 Words

In Literaria, mathematician Alex dreamed of writing a bestseller novel. He believed math could guide him in crafting a compelling story. Alex analyzed successful novels, identifying patterns and common elements.

He discovered that most bestsellers followed a specific formula: a strong opening, rising tension, a climax, and a satisfying resolution. Using this formula, Alex plotted his novel.

Alex also applied statistical analysis to his writing, studying word frequency and sentence length. He used algorithms to predict the emotional impact of plot twists.

When Alex published his book, it became an instant bestseller. Critics praised its perfect pacing and emotional depth.

50 Words

In Literaria, mathematician Alex used math to write a bestseller. He analyzed successful novels, identified patterns, and applied statistical analysis to his writing. His book became an instant hit, praised for its perfect pacing and emotional depth.

25 Words

Mathematician Alex used math to write a bestseller. His book became an instant hit, praised for its perfect pacing and emotional depth.

20 Words

Mathematician Alex used math to write a bestseller. His book became an instant hit, praised for its pacing and depth.

15 Words

Mathematician Alex used math to write a bestseller. It became an instant hit.

10 Words

Alex used math to write a bestseller. Instant hit.

7 Words

Alex's math bestseller became an instant hit.

5 Words

Math bestseller, instant hit.

4 Words

Math bestseller, hit.

3 Words

Math bestseller.

2 Words

Math bestseller.

1 Word

Bestseller.

0 Words

(Silence)

Explanation

The story illustrates how math can be used as a tool to create a bestseller novel. By compacting the story, we distill its essence, showing how the core message remains powerful even in its simplest form.

Detailed Comments

- **Clarity**: The story and its compacted versions maintain a clear message about the power of math in writing.

- **Dynamism**: The use of mathematical analysis adds a dynamic element to the narrative.
- **Efficiency**: The compacted versions efficiently convey the core message without losing its essence.
- **Effectiveness**: Each version effectively communicates the relationship between math and writing.
- **Realism**: The story is grounded in realistic applications of math in the creative process.

Series of Opinions

1. Math can be a powerful tool for creative writing.
2. Mathematical analysis can simplify complex tasks.
3. Compacting stories can reveal their core essence.
4. Math can guide practical ways to achieve literary success.
5. Practical math can inspire creativity.

Poetic Math Functions

1. **Clarity**: $f(x)=clear(x)$
2. **Dynamism**: $g(x)=dynamic(x)$
3. **Efficiency**: $h(x)=efficient(x)$

4. **Effectiveness**: i(x)=effective(x)
5. **Realism**: j(x)=realistic(x)

Combining these functions: $$S(x) = f(x) + g(x) + h(x) + i(x) + j(x)$$

This function S(x) represents an exceptional story that embodies clarity, dynamism, efficiency, effectiveness, and realism.

Freestyle Series of Ideas

1. **Math in Plot Development**: Using algorithms to predict reader engagement at different plot points.
2. **Character Arcs**: Applying statistical models to ensure balanced character development.
3. **Emotional Impact**: Using data analysis to craft emotionally resonant scenes.
4. **Pacing**: Employing mathematical models to optimize the pacing of the story.
5. **Reader Feedback**: Analyzing reader feedback to refine and improve the narrative.

Interesting New Math Proof of Concepts

1. **Narrative Algorithms**: Developing algorithms that can predict the success of different narrative structures.

2. **Emotional Resonance Models**: Creating models that quantify the emotional impact of various plot elements.

3. **Character Development Metrics**: Establishing metrics to measure and balance character arcs.

4. **Pacing Optimization**: Using mathematical models to optimize the pacing of a story for maximum reader engagement.

5. **Feedback Analysis Tools**: Developing tools to analyze reader feedback and refine the narrative based on data-driven insights.

Microfiction Story (299 words)

In the city of Literaria, a writer named Maya struggled to craft a compelling novel. She turned to her friend Leo, a mathematician, for help. Leo believed that math could enhance literary devices and improve storytelling.

Leo explained, "Think of metaphors as equations. They compare two different things to reveal a deeper truth. For example, 'Time is a thief' is like saying Time=Thief. It conveys the idea that time steals moments from our lives."

Maya was intrigued. She began to see similes as mathematical comparisons, using "like" or "as" to draw parallels. "Her smile was like sunshine" became Smile≈Sunshine.

Leo continued, "Alliteration is like a sequence in math. It creates a pattern that pleases the ear. 'Peter Piper picked a peck of pickled peppers' is a series of repeated sounds, much like a repeating decimal."

Maya applied these concepts to her writing. She used hyperbole to exaggerate for effect, much like multiplying a number to emphasize its magnitude. "I've told you a million times" became Told×10⁶.

She also used parallelism, creating balanced sentences that mirrored each other, like symmetrical equations. "She loved to read, to write, and to dream" became Read=Write=Dream.

Maya's novel transformed. Her use of literary devices, enhanced by mathematical principles, created a captivating and harmonious narrative. When she published her book, it became a bestseller, praised for its innovative storytelling.

Maya's success proved that math and literature could work together to create powerful and engaging stories. Her novel became a testament to the beauty of combining numbers and words.

Compacting the Story

250 Words

In Literaria, writer Maya struggled to craft a compelling novel. She turned to her friend Leo, a mathematician, for help. Leo believed math could enhance literary devices and improve storytelling.

Leo explained, "Think of metaphors as equations. They compare two different things to reveal a deeper truth. For example, 'Time is a thief' is like saying Time=Thief."

Maya was intrigued. She began to see similes as mathematical comparisons, using "like" or "as" to draw parallels. "Her smile was like sunshine" became Smile≈Sunshine.

Leo continued, "Alliteration is like a sequence in math. It creates a pattern that pleases the ear. 'Peter Piper picked a peck of pickled peppers' is a series of repeated sounds, much like a repeating decimal."

Maya applied these concepts to her writing. She used hyperbole to exaggerate for effect, much like multiplying a number to emphasize its magnitude. "I've told you a million times" became Told×10⁶.

She also used parallelism, creating balanced sentences that mirrored each other, like symmetrical equations. "She loved to read, to write, and to dream" became Read=Write=Dream.

Maya's novel transformed. Her use of literary devices, enhanced by mathematical principles, created a captivating narrative. When she published her book, it became a bestseller, praised for its innovative storytelling.

Maya's success proved that math and literature could work together to create powerful stories.

200 Words

In Literaria, writer Maya struggled to craft a compelling novel. She turned to her friend Leo, a mathematician, for help. Leo believed math could enhance literary devices and improve storytelling.

Leo explained, "Think of metaphors as equations. For example, 'Time is a thief' is like saying Time=Thief."

Maya was intrigued. She began to see similes as mathematical comparisons. "Her smile was like sunshine" became Smile≈Sunshine.

Leo continued, "Alliteration is like a sequence in math. 'Peter Piper picked a peck of pickled peppers' is a series of repeated sounds, much like a repeating decimal."

Maya applied these concepts to her writing. She used hyperbole to exaggerate for effect, much like multiplying a number. "I've told you a million times" became Told×106.

She also used parallelism, creating balanced sentences that mirrored each other, like symmetrical equations. "She loved to read, to write, and to dream" became Read=Write=Dream.

Maya's novel transformed. Her use of literary devices, enhanced by mathematical principles, created a captivating narrative. When she published her book, it became a bestseller, praised for its innovative storytelling.

Maya's success proved that math and literature could work together.

150 Words

In Literaria, writer Maya struggled to craft a compelling novel. She turned to her friend Leo, a mathematician, for help. Leo believed math could enhance literary devices.

Leo explained, "Think of metaphors as equations. For example, 'Time is a thief' is like saying Time=Thief."

Maya was intrigued. She began to see similes as mathematical comparisons. "Her smile was like sunshine" became Smile≈Sunshine.

Leo continued, "Alliteration is like a sequence in math. 'Peter Piper picked a peck of pickled peppers' is a series of repeated sounds, much like a repeating decimal."

Maya applied these concepts to her writing. She used hyperbole to exaggerate for effect, much like multiplying a number. "I've told you a million times" became Told×106.

She also used parallelism, creating balanced sentences that mirrored each other, like symmetrical equations. "She loved to read, to write, and to dream" became Read=Write=Dream.

Maya's novel transformed. When she published her book, it became a bestseller, praised for its innovative storytelling.

100 Words

In Literaria, writer Maya struggled to craft a compelling novel. She turned to her friend Leo, a mathematician, for help. Leo believed math could enhance literary devices.

Leo explained, "Think of metaphors as equations. For example, 'Time is a thief' is like saying Time=Thief."

Maya was intrigued. She began to see similes as mathematical comparisons. "Her smile was like sunshine" became Smile≈Sunshine.

Maya applied these concepts to her writing. She used hyperbole, much like multiplying a number. "I've told you a million times" became Told×106.

Maya's novel transformed. It became a bestseller.

50 Words

In Literaria, writer Maya struggled to craft a compelling novel. She turned to her friend Leo, a mathematician, for help. Leo believed math could enhance literary devices. Maya applied these concepts to her writing. Her novel transformed and became a bestseller.

25 Words

Writer Maya struggled to craft a novel. Mathematician Leo helped her use math to enhance literary devices. Her novel became a bestseller.

20 Words

Writer Maya struggled. Mathematician Leo helped her use math to enhance literary devices. Her novel became a bestseller.

15 Words

Writer Maya struggled. Mathematician Leo helped. Her novel became a bestseller.

10 Words

Maya struggled. Leo helped. Her novel became a bestseller.

7 Words

Maya struggled. Leo helped. Bestseller novel.

5 Words

Maya's novel: math helped.

4 Words

Math helped bestseller.

3 Words

Math helped.

2 Words

Math helped.

1 Word

Math.

0 Words

(Silence)

Explanation

The story illustrates how math can be used to enhance literary devices and improve storytelling. By compacting the story, we distill its essence, showing how the core message remains powerful even in its simplest form.

Detailed Comments

- **Clarity**: The story and its compacted versions maintain a clear message about the power of math in writing.
- **Dynamism**: The use of mathematical principles adds a dynamic element to the narrative.
- **Efficiency**: The compacted versions efficiently convey the core message without losing its essence.
- **Effectiveness**: Each version effectively communicates the relationship between math and literary devices.
- **Realism**: The story is grounded in realistic applications of math in the creative process.

Series of Opinions

1. Math can be a powerful tool for enhancing literary devices.
2. Mathematical principles can simplify complex tasks.
3. Compacting stories can reveal their core essence.
4. Math can guide practical ways to achieve literary success.
5. Practical math can inspire creativity.

Poetic Math Functions

1. **Clarity**: $f(x)=clear(x)$
2. **Dynamism**: $g(x)=dynamic(x)$
3. **Efficiency**: $h(x)=efficient(x)$
4. **Effectiveness**: $i(x)=effective(x)$
5. **Realism**: \

***

The relationship of math to origin of the universe:

Microfiction Story (299 words)

In the beginning, there was nothing but a void. Then, a singularity emerged, a point of infinite density and energy.

This singularity, governed by the laws of mathematics, expanded in a cosmic event known as the Big Bang. From this explosion, the universe was born.

As the universe expanded, particles began to form, guided by mathematical principles. The forces of gravity, electromagnetism, and nuclear interactions shaped the cosmos. Stars ignited, galaxies formed, and planets coalesced, all following the elegant equations of physics.

On a small blue planet, life emerged. Over billions of years, simple organisms evolved into complex beings capable of thought and curiosity. Among them was a mathematician named Dr. Elena. She dedicated her life to understanding the universe through the lens of mathematics.

Dr. Elena marveled at the beauty of the equations that described the cosmos. She saw the Fibonacci sequence in the spirals of galaxies, the symmetry of fractals in the branching of trees, and the golden ratio in the proportions of living organisms. She realized that math was the language of the universe, a code that unlocked the secrets of existence.

One night, as she gazed at the stars, Dr. Elena had an epiphany. She understood that the same mathematical principles that governed the birth of the universe also guided the evolution of life. The universe and life were interconnected, bound by the same elegant equations.

Dr. Elena's discoveries inspired a new generation of scientists and mathematicians. They continued to explore the cosmos, uncovering new mysteries and deepening their understanding of the universe. Through math, they glimpsed the profound beauty and order that lay at the heart of existence.

Compacting the Story

250 Words

In the beginning, a singularity emerged, governed by the laws of mathematics. This singularity expanded in the Big Bang, birthing the universe. As the universe expanded, particles formed, guided by mathematical principles. Stars ignited, galaxies formed, and planets coalesced, all following the elegant equations of physics.

On a small blue planet, life emerged. Over billions of years, simple organisms evolved into complex beings. Among them was Dr. Elena, a mathematician dedicated to understanding the universe through math.

Dr. Elena marveled at the beauty of the equations that described the cosmos. She saw the Fibonacci sequence in galaxies, the symmetry of fractals in trees, and the golden ratio in living organisms. She realized that math was the language of the universe, unlocking the secrets of existence.

One night, as she gazed at the stars, Dr. Elena had an epiphany. She understood that the same mathematical principles that governed the birth of the universe also guided the evolution of life. The universe and life were interconnected, bound by the same elegant equations.

Dr. Elena's discoveries inspired a new generation of scientists and mathematicians. They continued to explore the cosmos, uncovering new mysteries and deepening their understanding of the universe. Through math, they glimpsed the profound beauty and order at the heart of existence.

200 Words

In the beginning, a singularity emerged, governed by the laws of mathematics. This singularity expanded in the Big Bang, birthing the universe. As the universe expanded, particles formed, guided by mathematical principles. Stars ignited, galaxies formed, and planets coalesced, all following the elegant equations of physics.

On a small blue planet, life emerged. Over billions of years, simple organisms evolved into complex beings. Among them was Dr. Elena, a mathematician dedicated to understanding the universe through math.

Dr. Elena marveled at the beauty of the equations that described the cosmos. She saw the Fibonacci sequence in galaxies, the symmetry of fractals in trees, and the golden ratio in living organisms. She realized that math was the language of the universe, unlocking the secrets of existence.

One night, as she gazed at the stars, Dr. Elena had an epiphany. She understood that the same mathematical principles that governed the birth of the universe also guided the evolution of life. The universe and life were interconnected, bound by the same elegant equations.

Dr. Elena's discoveries inspired a new generation of scientists and mathematicians. They continued to explore the cosmos, uncovering new mysteries and deepening their understanding of the universe.

150 Words

In the beginning, a singularity emerged, governed by the laws of mathematics. This singularity expanded in the Big Bang, birthing the universe. As the universe expanded, particles formed, guided by mathematical principles. Stars ignited, galaxies formed, and planets coalesced, all following the elegant equations of physics.

On a small blue planet, life emerged. Among them was Dr. Elena, a mathematician dedicated to understanding the universe through math.

Dr. Elena marveled at the beauty of the equations that described the cosmos. She saw the Fibonacci sequence in galaxies, the symmetry of fractals in trees, and the golden ratio in living organisms. She realized that math was the language of the universe, unlocking the secrets of existence.

One night, as she gazed at the stars, Dr. Elena had an epiphany. She understood that the same mathematical principles that governed the birth of the universe also guided the evolution of life.

100 Words

In the beginning, a singularity emerged, governed by the laws of mathematics. This singularity expanded in the Big Bang, birthing the universe. As the universe expanded, particles formed, guided by mathematical principles. Stars ignited, galaxies formed, and planets coalesced.

On a small blue planet, life emerged. Among them was Dr. Elena, a mathematician dedicated to understanding the universe through math.

Dr. Elena marveled at the beauty of the equations that described the cosmos. She saw the Fibonacci sequence in galaxies and the golden ratio in living organisms. She realized that math was the language of the universe.

50 Words

In the beginning, a singularity expanded in the Big Bang, birthing the universe. Particles formed, guided by math. Stars ignited, galaxies formed, and life emerged. Dr. Elena, a mathematician, saw math's beauty in the cosmos and realized it was the universe's language.

25 Words

A singularity expanded, birthing the universe. Math guided particles, stars, galaxies, and life. Dr. Elena saw math's beauty and realized it was the universe's language.

20 Words

A singularity expanded, birthing the universe. Math guided particles, stars, galaxies, and life. Dr. Elena saw math's beauty.

15 Words

A singularity expanded, birthing the universe. Math guided particles, stars, galaxies, and life.

10 Words

A singularity expanded. Math guided particles, stars, galaxies, life.

7 Words

Math guided particles, stars, galaxies, life.

5 Words

Math guided the universe.

4 Words

Math guided universe.

3 Words

Math guided.

2 Words

Math guided.

1 Word

Math.

0 Words

(Silence)

CHAPTER 20

Highlights the relationship of math to finding a new practical series of idea for a fresh sustainable start as well as those ways and means do and find wealth and easily grow and sustainable system then build a truly great nation:

Microfiction Story (299 words)

In the heart of a struggling nation, a mathematician named Elena sought to find a new path to prosperity. She believed that math could provide practical solutions for a fresh, sustainable start. Elena gathered a team of experts in various fields and began to develop a series of ideas based on mathematical principles.

First, they focused on agriculture. Using algorithms, they optimized crop rotation and irrigation systems, ensuring maximum yield with minimal resources. This approach not only increased food production but also preserved the environment.

Next, they tackled energy. By applying mathematical models, they designed efficient renewable energy systems, harnessing solar and wind power to reduce the nation's reliance on fossil fuels. This shift not only provided clean energy but also created jobs and stimulated the economy.

Elena's team also addressed education. They developed a curriculum that integrated math into everyday learning, teaching students practical skills for the future. This approach fostered innovation and creativity, preparing the next generation to tackle future challenges.

Finally, they focused on wealth distribution. Using statistical analysis, they created a fair tax system that ensured resources were allocated equitably. This system reduced poverty and promoted social harmony.

The nation's transformation was remarkable. Elena's mathematical approach to problem-solving created a sustainable system that fostered growth and prosperity. The nation thrived, becoming a beacon of innovation and sustainability.

Elena's story proved that math could be a powerful tool in building a truly great nation. Her vision and dedication inspired others to embrace mathematical principles in their own lives, creating a ripple effect of positive change.

Compacting the Story

250 Words

In a struggling nation, mathematician Elena sought a path to prosperity through math. She gathered experts and developed ideas based on mathematical principles.

First, they optimized agriculture using algorithms, increasing food production while preserving the environment. Next, they designed efficient renewable energy systems, harnessing solar and wind power to reduce reliance on fossil fuels, creating jobs and stimulating the economy.

Elena's team also revamped education, integrating math into everyday learning to foster innovation and creativity. They prepared the next generation to tackle future challenges.

Finally, they created a fair tax system using statistical analysis, ensuring equitable resource allocation. This system reduced poverty and promoted social harmony.

The nation's transformation was remarkable. Elena's mathematical approach created a sustainable system that fostered growth and prosperity. The nation thrived, becoming a beacon of innovation and sustainability.

Elena's story proved that math could be a powerful tool in building a truly great nation. Her vision inspired others to embrace mathematical principles, creating a ripple effect of positive change.

200 Words

In a struggling nation, mathematician Elena sought prosperity through math. She gathered experts and developed ideas based on mathematical principles.

First, they optimized agriculture using algorithms, increasing food production while preserving the environment. Next, they designed efficient renewable energy systems, harnessing solar and wind power to reduce reliance on fossil fuels, creating jobs and stimulating the economy.

Elena's team revamped education, integrating math into everyday learning to foster innovation and creativity. They prepared the next generation to tackle future challenges.

Finally, they created a fair tax system using statistical analysis, ensuring equitable resource allocation. This system reduced poverty and promoted social harmony.

The nation's transformation was remarkable. Elena's mathematical approach created a sustainable system that fostered growth and prosperity. The nation thrived, becoming a beacon of innovation and sustainability.

Elena's story proved that math could be a powerful tool in building a truly great nation. Her vision inspired others to embrace mathematical principles, creating a ripple effect of positive change.

150 Words

In a struggling nation, mathematician Elena sought prosperity through math. She gathered experts and developed ideas based on mathematical principles.

First, they optimized agriculture using algorithms, increasing food production while preserving the environment. Next, they designed efficient renewable energy systems, harnessing solar and wind power to reduce reliance on fossil fuels, creating jobs and stimulating the economy.

Elena's team revamped education, integrating math into everyday learning to foster innovation and creativity. They prepared the next generation to tackle future challenges.

Finally, they created a fair tax system using statistical analysis, ensuring equitable resource allocation. This system reduced poverty and promoted social harmony.

The nation's transformation was remarkable. Elena's mathematical approach created a sustainable system that fostered growth and prosperity. The nation thrived, becoming a beacon of innovation and sustainability.

Elena's story proved that math could build a truly great nation.

100 Words

In a struggling nation, mathematician Elena sought prosperity through math. She gathered experts and developed ideas based on mathematical principles.

They optimized agriculture using algorithms, increasing food production while preserving the environment. They designed renewable energy systems, harnessing solar and wind power, creating jobs and stimulating the economy.

Elena's team revamped education, integrating math into everyday learning. They created a fair tax system using statistical analysis, ensuring equitable resource allocation.

The nation's transformation was remarkable. Elena's mathematical approach created a sustainable system that fostered growth and prosperity. The nation thrived, becoming a beacon of innovation and sustainability.

50 Words

Mathematician Elena sought prosperity through math. Her team optimized agriculture, designed renewable energy systems, revamped education, and created a fair tax system. The nation's transformation was remarkable. Elena's mathematical approach created a sustainable system that fostered growth and prosperity. The nation thrived.

25 Words

Elena used math to optimize agriculture, design renewable energy, revamp education, and create a fair tax system. The nation thrived.

20 Words

Elena used math to optimize agriculture, design renewable energy, revamp education, and create a fair tax system.

15 Words

Elena used math to optimize agriculture, design renewable energy, and revamp education.

10 Words

Elena used math to optimize agriculture and energy.

7 Words

Elena used math to optimize agriculture.

5 Words

Elena optimized agriculture with math.

4 Words

Math optimized agriculture.

3 Words

Math optimized.

2 Words

Math optimized.

1 Word

Math.

0 Words

(Silence)

Explanation

The story illustrates how math can be used as a tool to find practical solutions for a fresh, sustainable start and build a truly great nation. By compacting the story, we distill its essence, showing how the core message remains powerful even in its simplest form.

Detailed Comments

- **Clarity**: The story and its compacted versions maintain a clear message about the power of math in creating sustainable systems.

- **Dynamism**: The use of mathematical principles adds a dynamic element to the narrative.
- **Efficiency**: The compacted versions efficiently convey the core message without losing its essence.
- **Effectiveness**: Each version effectively communicates the relationship between math and sustainable development.
- **Realism**: The story is grounded in realistic applications of math in nation-building.

Series of Opinions

1. Math can be a powerful tool for sustainable development.
2. Mathematical principles can simplify complex tasks.
3. Compacting stories can reveal their core essence.
4. Math can guide practical ways to achieve prosperity and sustainability.
5. Practical math can inspire innovation and growth.

Poetic Math Functions

1. **Clarity**: $f(x)=clear(x)$
2. **Dynamism**: $g(x)=dynamic(x)$

3. **Efficiency**: h(x)=efficient(x)
4. **Effectiveness**: i(x)=effective(x)
5. **Realism**: j(x)=realistic(x)

Combining these functions: $$ S(x) = f(x) + g(x) + h(x) + i(x) + j(x) $$

This function S(x) represents an exceptional story that embodies clarity, dynamism, efficiency, effectiveness, and realism.

CHAPTER 21

Highlights the relationship of math to finding a new practical series of idea for a fresh sustainable start for a new epic forest adventure

Microfiction Story (299 words)

In the heart of a dense forest, a mathematician named Theo sought a fresh start. He believed that math could guide him in creating a sustainable and adventurous life. Theo gathered a group of like-minded individuals and set out on an epic forest adventure.

Using mathematical principles, Theo and his team devised a series of practical ideas for sustainable living. They began by mapping the forest, using geometry to plot the best locations for shelters, water sources, and food supplies. This ensured they had everything they needed within reach.

Next, they applied algorithms to optimize their foraging and hunting strategies. By analyzing patterns in animal behavior and plant growth, they maximized their resources while minimizing their impact on the environment.

Theo also used math to design efficient energy systems. They harnessed solar power and built wind turbines, ensuring a steady supply of renewable energy. This allowed them to live comfortably without relying on external resources.

The team also focused on education. They developed a curriculum that integrated math into their daily activities, teaching everyone practical skills for survival and sustainability. This approach fostered a sense of community and innovation.

As they thrived in the forest, Theo's mathematical approach proved invaluable. Their sustainable lifestyle became a model for others seeking a fresh start. The forest adventure was not just about survival but about creating a harmonious and prosperous life.

Theo's story demonstrated that math could be a powerful tool in finding practical solutions for a sustainable future. His vision and dedication inspired others to embrace mathematical principles in their own lives, creating a ripple effect of positive change.

Compacting the Story

250 Words

In a dense forest, mathematician Theo sought a fresh start. He believed math could guide him in creating a sustainable and adventurous life. Theo gathered like-minded individuals for an epic forest adventure.

Using mathematical principles, Theo's team devised practical ideas for sustainable living. They mapped the forest using geometry to plot the best locations for shelters, water sources, and food supplies.

Next, they applied algorithms to optimize foraging and hunting strategies, analyzing patterns in animal behavior and plant growth. This maximized resources while minimizing environmental impact.

Theo designed efficient energy systems, harnessing solar power and building wind turbines for renewable energy. This allowed them to live comfortably without external resources.

The team also focused on education, integrating math into daily activities to teach practical survival and sustainability skills. This fostered community and innovation.

As they thrived in the forest, Theo's mathematical approach proved invaluable. Their sustainable lifestyle became a model for others seeking a fresh start. The forest adventure was about creating a harmonious and prosperous life.

Theo's story demonstrated that math could be a powerful tool in finding practical solutions for a sustainable future. His vision inspired others to embrace mathematical principles, creating a ripple effect of positive change.

200 Words

In a dense forest, mathematician Theo sought a fresh start. He believed math could guide him in creating a sustainable and adventurous life. Theo gathered like-minded individuals for an epic forest adventure.

Using mathematical principles, Theo's team devised practical ideas for sustainable living. They mapped the forest using geometry to plot the best locations for shelters, water sources, and food supplies.

Next, they applied algorithms to optimize foraging and hunting strategies, analyzing patterns in animal behavior and plant growth. This maximized resources while minimizing environmental impact.

Theo designed efficient energy systems, harnessing solar power and building wind turbines for renewable energy.

This allowed them to live comfortably without external resources.

The team also focused on education, integrating math into daily activities to teach practical survival and sustainability skills. This fostered community and innovation.

As they thrived in the forest, Theo's mathematical approach proved invaluable. Their sustainable lifestyle became a model for others seeking a fresh start. The forest adventure was about creating a harmonious and prosperous life.

Theo's story demonstrated that math could be a powerful tool in finding practical solutions for a sustainable future. His vision inspired others to embrace mathematical principles.

150 Words

In a dense forest, mathematician Theo sought a fresh start. He believed math could guide him in creating a sustainable and adventurous life. Theo gathered like-minded individuals for an epic forest adventure.

Using mathematical principles, Theo's team devised practical ideas for sustainable living. They mapped the forest using geometry to plot the best locations for shelters, water sources, and food supplies.

Next, they applied algorithms to optimize foraging and hunting strategies, analyzing patterns in animal behavior and plant growth. This maximized resources while minimizing environmental impact.

Theo designed efficient energy systems, harnessing solar power and building wind turbines for renewable energy. This allowed them to live comfortably without external resources.

The team also focused on education, integrating math into daily activities to teach practical survival and sustainability skills. This fostered community and innovation.

As they thrived in the forest, Theo's mathematical approach proved invaluable. Their sustainable lifestyle became a model for others seeking a fresh start.

100 Words

In a dense forest, mathematician Theo sought a fresh start. He believed math could guide him in creating a sustainable and adventurous life. Theo gathered like-minded individuals for an epic forest adventure.

Using mathematical principles, Theo's team mapped the forest, optimized foraging and hunting strategies, and designed efficient energy systems. They integrated math into daily activities to teach practical survival and sustainability skills.

As they thrived in the forest, Theo's mathematical approach proved invaluable. Their sustainable lifestyle became a model for others seeking a fresh start.

50 Words

In a dense forest, mathematician Theo sought a fresh start. Using math, his team mapped the forest, optimized resources, and designed energy systems. They integrated math into daily activities. Their sustainable lifestyle became a model for others.

25 Words

Mathematician Theo used math to create a sustainable forest life. His team's approach became a model for others.

20 Words

Mathematician Theo used math to create a sustainable forest life. His team's approach inspired others.

15 Words

Theo used math to create a sustainable forest life. His approach inspired others.

10 Words

Theo used math for a sustainable forest life.

7 Words

Theo's math created a sustainable life.

5 Words

Theo's math inspired sustainability.

4 Words

Math inspired sustainability.

3 Words

Math inspired.

2 Words

Math inspired.

1 Word

Math.

0 Words

(Silence)

Explanation

The story illustrates how math can be used as a tool to find practical solutions for a fresh, sustainable start in a forest adventure. By compacting the story, we distill its essence, showing how the core message remains powerful even in its simplest form.

Detailed Comments

- **Clarity**: The story and its compacted versions maintain a clear message about the power of math in creating sustainable systems.
- **Dynamism**: The use of mathematical principles adds a dynamic element to the narrative.
- **Efficiency**: The compacted versions efficiently convey the core message without losing its essence.
- **Effectiveness**: Each version effectively communicates the relationship between math and sustainable development.
- **Realism**: The story is grounded in realistic applications of math in a forest adventure.

Series of Opinions

1. Math can be a powerful tool for sustainable development.
2. Mathematical principles can simplify complex tasks.
3. Compacting stories can reveal their core essence.

4. Math can guide practical ways to achieve sustainability.

5. Practical math can inspire innovation and growth.

Poetic Math Functions

1. **Clarity**: f(x)=clear(x)

2. **Dynamism**: \

***

Or another example: highlights the relationship of math to finding a new practical series of idea for a fresh sustainable start for a new epic forest adventure.

***

Or another example, highlights the relationship of math to getting a divorce:

Microfiction Story (299 words)

In the city of Numeria, a couple named Alice and Ben were facing the end of their marriage. Both were mathematicians, and they decided to approach their divorce with the same logic and precision they applied to their work.

Alice and Ben sat down with a whiteboard and began to list their assets and debts. They created equations to divide their property fairly, ensuring that each received an equal share. They used Venn diagrams to determine custody arrangements for their children, ensuring that both parents had ample time with them.

As they worked through the numbers, they realized that their relationship had become a series of calculations. They had lost the emotional connection that once bound them together. Alice and Ben decided to use their mathematical skills to find a way to part amicably.

They created a formula for their divorce: $D=A+B2$, where D represented the division of their assets, A represented Alice's share, and B represented Ben's share. They applied this formula to every aspect of their separation, from finances to parenting.

The process was not without its challenges, but Alice and Ben found solace in the logic and fairness of their approach. They realized that while their marriage had ended, their respect for each other remained intact.

In the end, Alice and Ben's divorce was finalized with minimal conflict. They continued to co-parent their children and even collaborated on mathematical research. Their story became a testament to the power of math in navigating life's most difficult challenges.

Compacting the Story

250 Words

In Numeria, mathematicians Alice and Ben faced the end of their marriage. They decided to approach their divorce with the same logic and precision they applied to their work.

Alice and Ben listed their assets and debts, creating equations to divide their property fairly. They used Venn diagrams to determine custody arrangements for their children, ensuring both parents had ample time with them.

As they worked through the numbers, they realized their relationship had become a series of calculations. They had lost the emotional connection that once bound them together. Alice and Ben decided to use their mathematical skills to part amicably.

They created a formula for their divorce: $D=A+B2$, where D represented the division of their assets, A represented Alice's share, and B represented Ben's share. They applied this formula to every aspect of their separation, from finances to parenting.

The process was challenging, but Alice and Ben found solace in the logic and fairness of their approach. They realized that while their marriage had ended, their respect for each other remained intact.

In the end, Alice and Ben's divorce was finalized with minimal conflict. They continued to co-parent their children and even collaborated on mathematical research. Their story became a testament to the power of math in navigating life's most difficult challenges.

200 Words

In Numeria, mathematicians Alice and Ben faced the end of their marriage. They decided to approach their divorce with the same logic and precision they applied to their work.

Alice and Ben listed their assets and debts, creating equations to divide their property fairly. They used Venn diagrams to determine custody arrangements for their children, ensuring both parents had ample time with them.

As they worked through the numbers, they realized their relationship had become a series of calculations. They had lost the emotional connection that once bound them together. Alice and Ben decided to use their mathematical skills to part amicably.

They created a formula for their divorce: $D=A+B2$, where D represented the division of their assets, A represented

Alice's share, and B represented Ben's share. They applied this formula to every aspect of their separation.

The process was challenging, but Alice and Ben found solace in the logic and fairness of their approach. They realized that while their marriage had ended, their respect for each other remained intact.

In the end, Alice and Ben's divorce was finalized with minimal conflict. They continued to co-parent their children and even collaborated on mathematical research.

150 Words

In Numeria, mathematicians Alice and Ben faced the end of their marriage. They decided to approach their divorce with the same logic and precision they applied to their work.

Alice and Ben listed their assets and debts, creating equations to divide their property fairly. They used Venn diagrams to determine custody arrangements for their children.

As they worked through the numbers, they realized their relationship had become a series of calculations. They had lost the emotional connection that once bound them together. Alice and Ben decided to use their mathematical skills to part amicably.

They created a formula for their divorce: D=A+B2. They applied this formula to every aspect of their separation.

The process was challenging, but Alice and Ben found solace in the logic and fairness of their approach. They realized that while their marriage had ended, their respect for each other remained intact.

100 Words

In Numeria, mathematicians Alice and Ben faced the end of their marriage. They decided to approach their divorce with the same logic and precision they applied to their work.

Alice and Ben listed their assets and debts, creating equations to divide their property fairly. They used Venn diagrams to determine custody arrangements for their children.

They created a formula for their divorce: D=A+B2. They applied this formula to every aspect of their separation.

The process was challenging, but Alice and Ben found solace in the logic and fairness of their approach. Their respect for each other remained intact.

50 Words

In Numeria, mathematicians Alice and Ben faced divorce. They used math to divide assets and determine custody.

They created a formula: $D=A+B2$. The process was challenging, but they found solace in the logic and fairness. Their respect for each other remained intact.

25 Words

Mathematicians Alice and Ben faced divorce. They used math to divide assets and determine custody. Their respect for each other remained intact.

20 Words

Mathematicians Alice and Ben faced divorce. They used math to divide assets and determine custody. Respect remained.

15 Words

Alice and Ben faced divorce. They used math to divide assets. Respect remained.

10 Words

Alice and Ben used math to divide assets.

7 Words

Alice and Ben used math. Respect remained.

5 Words

Math divided assets. Respect remained.

4 Words

Math divided assets.

3 Words

Math divided.

2 Words

Math divided.

1 Word

Divided.

0 Words

(Silence)

Explanation

The story illustrates how math can be used as a tool to navigate the complexities of divorce. By compacting the story, we distill its essence, showing how the core message remains powerful even in its simplest form.

Detailed Comments

- **Clarity**: The story and its compacted versions maintain a clear message about the power of math in navigating divorce.

- **Dynamism**: The use of mathematical principles adds a dynamic element to the narrative.
- **Efficiency**: The compacted versions efficiently convey the core message without losing its essence.
- **Effectiveness**: Each version effectively communicates the relationship between math and divorce.
- **Realism**: The story is grounded in realistic applications of math in personal relationships.

Series of Opinions

1. Math can be a powerful tool for navigating personal challenges.
2. Mathematical principles can simplify complex tasks.
3. Compacting stories can reveal their core essence.
4. Math can guide practical ways to achieve fairness and respect.
5. Practical math can inspire harmony.

Poetic Math Functions

1. **Clarity**: $f(x)=\text{clear}(x)$
2. **Dynamism**: $g(x)=\text{dynamic}(x)$

3. **Efficiency**: h(x)=efficient(x)
4. **Effectiveness**: i(x)=effective(x)
5. **Realism**: j(x)=realistic(x)

Combining these functions: $$ S(x) = f(x) + g(x) + h(x) + i(x) + j(x) $$

CHAPTER 22

Find a great treasure map and prep then go on that fresh adventure:

Microfiction Story (299 words)

In the quaint town of Eldoria, a young adventurer named Leo stumbled upon an ancient map hidden in his grandfather's attic. The map promised a great treasure buried deep within the Whispering Woods. Excited, Leo gathered his friends, Maya and Ben, to embark on this fresh adventure.

They spent days preparing for the journey. Leo used his cartography skills to plot their route, while Maya, a botanist, identified the plants they might encounter. Ben, an engineer, designed tools to help them navigate the dense forest and dig for the treasure.

With their backpacks filled with supplies, they set off at dawn. The Whispering Woods were as enchanting as they were treacherous. Using the map, they followed a series of mathematical clues that led them through winding paths and hidden clearings. Each clue required them to solve puzzles and riddles, testing their wits and teamwork.

After days of trekking, they reached a clearing marked with an ancient symbol. Ben used his tools to dig at the indicated spot, and soon they unearthed a chest filled with gold coins, jewels, and ancient artifacts. The treasure was more than they had ever imagined.

Their adventure didn't just bring them wealth; it strengthened their friendship and taught them the value of teamwork and perseverance. They returned to Eldoria as heroes, their story inspiring others to seek their own adventures.

Compacting the Story

250 Words

In Eldoria, young adventurer Leo found an ancient map in his grandfather's attic, promising a great treasure in the Whispering Woods. He gathered his friends, Maya and Ben, to embark on the adventure.

They prepared meticulously. Leo plotted their route, Maya identified plants, and Ben designed tools for navigation and digging.

With supplies packed, they set off at dawn. The Whispering Woods were enchanting yet treacherous. Using the map, they followed mathematical clues through winding paths and hidden clearings, solving puzzles and riddles.

After days of trekking, they reached a clearing marked with an ancient symbol. Ben dug at the spot, unearthing a chest filled with gold coins, jewels, and artifacts.

The adventure brought them wealth and strengthened their friendship, teaching them the value of teamwork and perseverance. They returned to Eldoria as heroes, inspiring others to seek their own adventures.

200 Words

In Eldoria, adventurer Leo found an ancient map in his grandfather's attic, promising treasure in the Whispering Woods. He gathered friends Maya and Ben for the adventure.

They prepared meticulously. Leo plotted their route, Maya identified plants, and Ben designed tools for navigation and digging.

With supplies packed, they set off at dawn. The Whispering Woods were enchanting yet treacherous. Using the map, they followed mathematical clues through winding paths and hidden clearings, solving puzzles and riddles.

After days of trekking, they reached a clearing marked with an ancient symbol. Ben dug at the spot, unearthing a chest filled with gold coins, jewels, and artifacts.

The adventure brought them wealth and strengthened their friendship, teaching them the value of teamwork and perseverance. They returned to Eldoria as heroes, inspiring others to seek their own adventures.

150 Words

In Eldoria, adventurer Leo found an ancient map in his grandfather's attic, promising treasure in the Whispering Woods. He gathered friends Maya and Ben for the adventure.

They prepared meticulously. Leo plotted their route, Maya identified plants, and Ben designed tools for navigation and digging.

With supplies packed, they set off at dawn. The Whispering Woods were enchanting yet treacherous. Using the map, they followed mathematical clues through winding paths and hidden clearings, solving puzzles and riddles.

After days of trekking, they reached a clearing marked with an ancient symbol. Ben dug at the spot, unearthing a chest filled with gold coins, jewels, and artifacts.

The adventure brought them wealth and strengthened their friendship, teaching them the value of teamwork and perseverance. They returned to Eldoria as heroes.

100 Words

In Eldoria, adventurer Leo found an ancient map in his grandfather's attic, promising treasure in the Whispering Woods. He gathered friends Maya and Ben for the adventure.

They prepared meticulously and set off at dawn. The Whispering Woods were enchanting yet treacherous. Using the map, they followed mathematical clues, solving puzzles and riddles.

After days of trekking, they reached a clearing marked with an ancient symbol. Ben dug at the spot, unearthing a chest filled with gold coins, jewels, and artifacts.

The adventure brought them wealth and strengthened their friendship. They returned to Eldoria as heroes.

50 Words

In Eldoria, adventurer Leo found an ancient map promising treasure. With friends Maya and Ben, he followed mathematical clues through the Whispering Woods. They unearthed a chest of gold and jewels, strengthening their friendship. They returned as heroes.

25 Words

Leo found a treasure map. With friends, he followed clues through the woods, unearthed gold and jewels, and returned as heroes.

20 Words

Leo found a treasure map. With friends, he unearthed gold and jewels, returning as heroes.

15 Words

Leo found a map. With friends, he unearthed treasure, returning as heroes.

10 Words

Leo found a map, unearthed treasure, returned as heroes.

7 Words

Leo found treasure, returned as heroes.

5 Words

Found treasure, returned heroes.

4 Words

Found treasure, heroes.

3 Words

Treasure, heroes.

2 Words

Found treasure.

1 Word

Treasure.

0 Words

(Silence)

Explanation

The story illustrates the excitement and teamwork involved in finding a treasure map and embarking on an adventure. By compacting the story, we distill its essence, showing how the core message remains powerful even in its simplest form.

Detailed Comments

- **Clarity**: The story and its compacted versions maintain a clear message about the adventure and teamwork.
- **Dynamism**: The use of mathematical clues adds a dynamic element to the narrative.

- **Efficiency**: The compacted versions efficiently convey the core message without losing its essence.
- **Effectiveness**: Each version effectively communicates the relationship between math and adventure.
- **Realism**: The story is grounded in realistic applications of math in problem-solving.

Series of Opinions

1. Math can be a powerful tool for adventure and problem-solving.
2. Mathematical clues can simplify complex tasks.
3. Compacting stories can reveal their core essence.
4. Math can guide practical ways to achieve success and teamwork.
5. Practical math can inspire creativity and exploration.

Poetic Math Functions

1. **Clarity**: $f(x)=\text{clear}(x)$
2. **Dynamism**: $g(x)=\text{dynamic}(x)$
3. **Efficiency**: $h(x)=\text{efficient}(x)$

4. **Effectiveness**: i(x)=effective(x)

5. **Realism**: j(x)=realistic(x)

Combining these functions: $$S(x) = f(x) + g(x) + h(x) + i(x) + j(x)$$

This function S(x) represents an exceptional story that embodies clarity, dynamism, efficiency, effectiveness, and realism.

***

Highlight the relationship of math to finding a new practical series of idea for shopping for food that helps lose weight, be heathier and get a fresh sustainable start:

Microfiction Story (299 words)

In the bustling city of Metropolis, Emma, a data scientist, was on a mission to revolutionize grocery shopping. She believed that math could help people make healthier choices, lose weight, and embrace sustainability. Armed with her algorithms, she set out to create a new shopping system.

Emma's idea was simple yet powerful: use mathematical models to optimize grocery lists. She developed an app that analyzed users' dietary needs, weight loss goals, and environmental impact preferences. The app used linear programming to suggest the best combination of foods that met these criteria.

One day, Emma met John, a busy professional struggling with his weight and health. Intrigued by her app, he decided to give it a try. The app recommended a balanced mix of fresh produce, lean proteins, and whole grains, all sourced from local farms to minimize environmental impact.

John followed the app's suggestions and noticed significant changes. He lost weight, felt more energetic, and even saved money by avoiding processed foods. The app's sustainability feature also made him feel good about his choices, knowing he was contributing to a healthier planet.

Word of Emma's app spread, and soon, people all over Metropolis were using it. The city's grocery stores adapted, stocking more local and sustainable options. Emma's mathematical approach had not only transformed individual lives but also the community's shopping habits.

Emma's success proved that math could be a powerful tool for positive change. By combining data science with practical applications, she had created a system that helped people lead healthier, more sustainable lives. And in the process, she had shown that even the most complex problems could be solved with a little creativity and a lot of math.

Compact Story (250 words)

In Metropolis, Emma, a data scientist, aimed to revolutionize grocery shopping using math. She created an app that optimized grocery lists based on dietary needs, weight loss goals, and environmental impact preferences.

Emma's app used linear programming to suggest the best combination of foods. John, a busy professional struggling with his weight, decided to try it. The app recommended a balanced mix of fresh produce, lean proteins, and whole grains, sourced from local farms.

John followed the app's suggestions and noticed significant changes. He lost weight, felt more energetic, and saved money by avoiding processed foods. The app's sustainability feature also made him feel good about his choices.

Word of Emma's app spread, and soon, people all over Metropolis were using it. Grocery stores adapted, stocking more local and sustainable options. Emma's mathematical approach transformed individual lives and the community's shopping habits.

Emma's success proved that math could be a powerful tool for positive change. By combining data science with practical applications, she created a system that helped people lead healthier, more sustainable lives. She showed that even complex problems could be solved with creativity and math.

Compact Story (200 words)

In Metropolis, Emma, a data scientist, aimed to revolutionize grocery shopping using math. She created an app that optimized grocery lists based on dietary needs, weight loss goals, and environmental impact preferences.

Emma's app used linear programming to suggest the best combination of foods. John, a busy professional struggling with his weight, decided to try it. The app recommended a balanced mix of fresh produce, lean proteins, and whole grains, sourced from local farms.

John followed the app's suggestions and noticed significant changes. He lost weight, felt more energetic, and saved money by avoiding processed foods. The app's sustainability feature also made him feel good about his choices.

Word of Emma's app spread, and soon, people all over Metropolis were using it. Grocery stores adapted, stocking more local and sustainable options. Emma's mathematical approach transformed individual lives and the community's shopping habits.

Emma's success proved that math could be a powerful tool for positive change. By combining data science with practical applications, she created a system that helped people lead healthier, more sustainable lives.

Compact Story (150 words)

In Metropolis, Emma, a data scientist, created an app to revolutionize grocery shopping using math. The app optimized grocery lists based on dietary needs, weight loss goals, and environmental impact preferences.

John, a busy professional struggling with his weight, tried the app. It recommended a balanced mix of fresh produce, lean proteins, and whole grains, sourced from local farms.

John followed the app's suggestions and noticed significant changes. He lost weight, felt more energetic, and saved money by avoiding processed foods. The app's sustainability feature also made him feel good about his choices.

Word of Emma's app spread, and soon, people all over Metropolis were using it. Grocery stores adapted, stocking more local and sustainable options. Emma's mathematical approach transformed individual lives and the community's shopping habits.

Emma's success proved that math could be a powerful tool for positive change.

Compact Story (100 words)

In Metropolis, Emma, a data scientist, created an app to revolutionize grocery shopping using math. The app optimized grocery lists based on dietary needs, weight loss goals, and environmental impact preferences.

John, struggling with his weight, tried the app. It recommended a balanced mix of fresh produce, lean proteins, and whole grains, sourced from local farms.

John followed the app's suggestions and noticed significant changes. He lost weight, felt more energetic, and saved money. The app's sustainability feature also made him feel good about his choices.

Emma's mathematical approach transformed individual lives and the community's shopping habits.

Compact Story (50 words)

Emma, a data scientist, created an app to optimize grocery lists using math. John, struggling with his weight, tried it.

The app recommended balanced, sustainable foods. John lost weight, felt better, and saved money. Emma's app transformed shopping habits in Metropolis, proving math's power for positive change.

Compact Story (25 words)

Emma's app used math to optimize grocery lists. John lost weight, felt better, and saved money. The app transformed Metropolis's shopping habits, proving math's power.

Compact Story (20 words)

Emma's app used math for grocery lists. John lost weight, felt better, saved money. The app transformed Metropolis's shopping habits.

Compact Story (15 words)

Emma's app used math for groceries. John lost weight, felt better. The app transformed shopping habits.

Compact Story (10 words)

***

The next example highlights the relationship of math to finding a new practical way to pick top quality growth stocks to invest:

Microfiction Story (299 words)

In the bustling city of Financia, a young mathematician named Clara sought to find a new practical way to pick top-quality growth stocks to invest in. She believed that math could unlock the secrets of the stock market. Clara gathered a team of financial analysts and data scientists to develop a mathematical model for stock selection.

They began by analyzing historical stock data, identifying patterns and trends. Clara used statistical analysis to determine key indicators of growth, such as revenue growth rate, profit margins, and return on equity. She created a formula that combined these indicators into a single score, which she called the Growth Potential Index (GPI).

Clara's team tested the GPI on past stock performance, and the results were promising. Stocks with high GPI scores consistently outperformed the market. Encouraged by their success, they applied the model to current stock data and identified a list of top-quality growth stocks.

Clara and her team presented their findings to a group of investors. The investors were impressed by the mathematical rigor and practical application of the GPI. They decided to invest in the recommended stocks, and within months, their portfolios saw significant gains.

The success of the GPI model spread quickly, and Clara's approach became a standard in the financial industry. Her story proved that math could be a powerful tool in making informed investment decisions. Clara's innovative method not only brought wealth to investors but also inspired others to embrace mathematical principles in their financial strategies.

Compacting the Story

250 Words

In Financia, mathematician Clara sought a new way to pick top-quality growth stocks. She believed math could unlock the stock market's secrets. Clara gathered financial analysts and data scientists to develop a mathematical model for stock selection.

They analyzed historical stock data, identifying patterns and trends. Clara used statistical analysis to determine key growth indicators, such as revenue growth rate, profit margins, and return on equity. She created a formula called the Growth Potential Index (GPI).

Clara's team tested the GPI on past stock performance, and the results were promising. Stocks with high GPI scores consistently outperformed the market. Encouraged by their success, they applied the model to current stock data and identified top-quality growth stocks.

Clara and her team presented their findings to investors, who were impressed by the GPI's mathematical rigor and practical application. They invested in the recommended stocks, and within months, their portfolios saw significant gains.

The success of the GPI model spread quickly, becoming a standard in the financial industry. Clara's story proved that math could be a powerful tool in making informed investment decisions. Her innovative method brought wealth to investors and inspired others to embrace mathematical principles in their financial strategies.

200 Words

In Financia, mathematician Clara sought a new way to pick top-quality growth stocks. She believed math could unlock the stock market's secrets. Clara gathered financial analysts and data scientists to develop a mathematical model for stock selection.

They analyzed historical stock data, identifying patterns and trends. Clara used statistical analysis to determine key growth indicators, such as revenue growth rate, profit margins, and return on equity. She created a formula called the Growth Potential Index (GPI).

Clara's team tested the GPI on past stock performance, and the results were promising. Stocks with high GPI scores consistently outperformed the market. Encouraged by their success, they applied the model to current stock data and identified top-quality growth stocks.

Clara and her team presented their findings to investors, who were impressed by the GPI's mathematical rigor and practical application. They invested in the recommended stocks, and within months, their portfolios saw significant gains.

The success of the GPI model spread quickly, becoming a standard in the financial industry. Clara's story proved that math could be a powerful tool in making informed investment decisions. Her innovative method brought wealth to investors and inspired others to embrace mathematical principles in their financial strategies.

150 Words

In Financia, mathematician Clara sought a new way to pick top-quality growth stocks. She believed math could unlock the stock market's secrets. Clara gathered financial analysts and data scientists to develop a mathematical model for stock selection.

They analyzed historical stock data, identifying patterns and trends. Clara used statistical analysis to determine key growth indicators, such as revenue growth rate, profit margins, and return on equity. She created a formula called the Growth Potential Index (GPI).

Clara's team tested the GPI on past stock performance, and the results were promising. Stocks with high GPI scores consistently outperformed the market. Encouraged by their success, they applied the model to current stock data and identified top-quality growth stocks.

Clara and her team presented their findings to investors, who were impressed by the GPI's mathematical rigor and practical application. They invested in the recommended stocks, and within months, their portfolios saw significant gains.

100 Words

In Financia, mathematician Clara sought a new way to pick top-quality growth stocks. She believed math could unlock the stock market's secrets. Clara gathered financial analysts and data scientists to develop a mathematical model for stock selection.

They analyzed historical stock data, identifying patterns and trends. Clara used statistical analysis to determine key growth indicators, creating the Growth Potential Index (GPI).

Clara's team tested the GPI, and the results were promising. Stocks with high GPI scores consistently outperformed the market. Investors were impressed and invested in the recommended stocks, seeing significant gains.

50 Words

In Financia, mathematician Clara developed the Growth Potential Index (GPI) to pick top-quality growth stocks. The GPI used statistical analysis to identify key growth indicators. Investors were impressed and invested in the recommended stocks, seeing significant gains.

25 Words

Mathematician Clara developed the Growth Potential Index (GPI) to pick growth stocks. Investors saw significant gains.

20 Words

Mathematician Clara developed the Growth Potential Index (GPI). Investors saw gains.

15 Words

Clara developed the Growth Potential Index. Investors saw gains.

10 Words

Clara's Growth Potential Index led to gains.

7 Words

Clara's GPI led to gains.

5 Words

GPI led to gains.

4 Words

GPI led gains.

3 Words

GPI gains.

2 Words

GPI gains.

1 Word

Gains.

0 Words

(Silence)

Explanation

The story illustrates how math can be used to develop a practical method for picking top-quality growth stocks. By compacting the story, we distill its essence, showing how the core message remains powerful even in its simplest form.

Detailed Comments

- **Clarity**: The story and its compacted versions maintain a clear message about the power of math in stock selection.
- **Dynamism**: The use of mathematical principles adds a dynamic element to the narrative.
- **Efficiency**: The compacted versions efficiently convey the core message without losing its essence.
- **Effectiveness**: Each version effectively communicates the relationship between math and investment.
- **Realism**: The story is grounded in realistic applications of math in financial strategies.

Series of Opinions

1. Math can be a powerful tool for investment strategies.

2. Mathematical principles can simplify complex tasks.
3. Compacting stories can reveal their core essence.
4. Math can guide practical ways to achieve financial success.
5. Practical math can inspire confidence in investment decisions.

Sonnets

Sonnet 1: The Growth Potential Index

In Financia, where numbers rule the day, A mathematician sought a novel way, To pick the stocks that promised growth and gain, And free investors from financial pain.

With data, trends, and patterns she did play, Creating formulas to light the way. The Growth Potential Index was her guide, A tool to help the market's waves to ride.

Investors marveled at her keen insight, Their portfolios soared to greater height. With math and logic, Clara led the way, To wealth and fortune, bright as light of day.

Her story proves that numbers hold the key, To financial success and prosperity.

Sonnet 2: Clara's Mathematical Quest

In Financia, Clara's quest began, To find a way to help the common man. With math and data, she did seek to find, A method to bring peace to troubled mind.

She gathered experts, analysts, and more, To build a model, solid to the core. The Growth Potential Index was their aim, A formula to bring investors fame.

With careful thought and rigorous design, They crafted tools to help the stocks align. Investors saw their fortunes start to rise, As Clara's math brought wealth before their eyes.

Her tale reminds us all of math's great power, To turn the tide and bring a brighter hour.

Epic Poem: The Quest for Growth

In Financia, a city bright and grand, A mathematician took a daring stand. Clara was her name, a mind so keen, She sought to find a way to intervene.

The stock market, a place of risk and fear, She aimed to bring a method crystal clear. With analysts and data by her side, She ventured forth, her knowledge as her guide.

They studied trends and patterns of the past, To build a model that would truly last. The Growth Potential Index was their goal, A tool to help investors reach their role.

Through trials and tests, they forged a path so bright, Their formula a beacon in the night. Investors marveled at the gains they made, As Clara's math brought wealth that would not fade.

Her story spread, a legend far and wide, Of how math's power turned the market's tide. In Financia, her name would ever be, A symbol of success and victory.

CHAPTER 23

The relationship of math to finding a sustainable happiness:

Microfiction Story (299 words)

In the serene town of Harmonyville, a mathematician named Emma sought to find a formula for sustainable happiness. She believed that math could provide a practical approach to achieving lasting contentment. Emma began by analyzing various aspects of life, such as relationships, work, health, and leisure.

She created a mathematical model that assigned values to different activities and their impact on happiness. Emma called it the Happiness Equation.

Emma tested her equation by applying it to her own life. She balanced her time between nurturing relationships, pursuing meaningful work, maintaining her health, and enjoying leisure activities. She tracked her happiness levels over several months and found that the equation accurately predicted her overall well-being.

Encouraged by her success, Emma shared her findings with the community. People began to use the Happiness

Equation to guide their daily lives, making adjustments to achieve a balanced and fulfilling lifestyle. The town of Harmonyville thrived, with residents experiencing increased happiness and a stronger sense of community.

Emma's story proved that math could be a powerful tool in finding sustainable happiness. Her innovative approach inspired others to embrace mathematical principles in their pursuit of contentment. Harmonyville became a beacon of happiness, demonstrating that a balanced life, guided by math, could lead to lasting joy.

Compacting the Story

250 Words

InHarmonyville, mathematician Emma sought a formula for sustainable happiness. She believed math could provide a practical approach to lasting contentment. Emma analyzed aspects of life, such as relationships, work, health, and leisure. She created the Happiness Equation.

Emma tested her equation by applying it to her own life. She balanced her time between relationships, work, health, and leisure. She tracked her happiness levels over several months and found the equation accurately predicted her well-being.

Encouraged by her success, Emma shared her findings with the community. People used the Happiness Equation to guide their lives, achieving a balanced and fulfilling lifestyle. Harmonyville thrived, with residents experiencing increased happiness and a stronger sense of community.

Emma's story proved math could be a powerful tool in finding sustainable happiness. Her approach inspired others to embrace mathematical principles in their pursuit of contentment. Harmonyville became a beacon of happiness, demonstrating that a balanced life, guided by math, could lead to lasting joy.

200 Words

InHarmonyville, mathematician Emma sought a formula for sustainable happiness. She believed math could provide a practical approach to lasting contentment. Emma analyzed aspects of life, such as relationships, work, health, and leisure.

She created the Happiness Equation.

Emma tested her equation by applying it to her own life. She balanced her time between relationships, work, health, and leisure. She tracked her happiness levels over several months and found the equation accurately predicted her well-being.

Encouraged by her success, Emma shared her findings with the community. People used the Happiness Equation to guide their lives, achieving a balanced and fulfilling lifestyle. Harmonyville thrived, with residents experiencing increased happiness and a stronger sense of community.

Emma's story proved math could be a powerful tool in finding sustainable happiness. Her approach inspired others to embrace mathematical principles in their pursuit of contentment.

150 Words

In Harmonyville, mathematician Emma sought a formula for sustainable happiness. She believed math could provide a practical approach to lasting contentment. Emma analyzed aspects of life, such as relationships, work, health, and leisure.

She created the Happiness Equation.

Emma tested her equation by applying it to her own life. She balanced her time between relationships, work, health, and leisure. She tracked her happiness levels over several months and found the equation accurately predicted her well-being.

Encouraged by her success, Emma shared her findings with the community. People used the Happiness Equation to guide their lives, achieving a balanced and fulfilling lifestyle. Harmonyville thrived, with residents experiencing increased happiness and a stronger sense of community.

100 Words

In Harmonyville, mathematician Emma sought a formula for sustainable happiness. She believed math could provide a practical approach to lasting contentment. Emma analyzed aspects of life, such as relationships, work, health, and leisure.

She created the Happiness Equation.

Emma tested her equation by applying it to her own life. She balanced her time between relationships, work, health, and leisure. She tracked her happiness levels and found the equation accurately predicted her well-being.

Encouraged by her success, Emma shared her findings with the community. Harmonyville thrived, with residents experiencing increased happiness.

50 Words

In Harmonyville, mathematician Emma sought a formula for sustainable happiness. She created the Happiness Equation. Emma tested it in her life and shared her findings. Harmonyville thrived, with residents experiencing increased happiness.

25 Words

Mathematician Emma created the Happiness Equation. Harmonyville thrived, with residents experiencing increased happiness.

20 Words

Mathematician Emma created the Happiness Equation. Harmonyville thrived, with residents experiencing increased happiness.

15 Words

Emma created the Happiness Equation. Harmonyville thrived, with residents experiencing happiness.

10 Words

Emma's Happiness Equation led to increased happiness in Harmonyville.

7 Words

Emma's equation increased happiness in Harmonyville.

5 Words

Equation increased happiness in Harmonyville.

4 Words

Equation increased happiness.

3 Words

Increased happiness.

2 Words

Happiness increased.

1 Word

Happiness.

0 Words

(Silence)

Explanation

The story illustrates how math can be used to find sustainable happiness. By compacting the story, we distill its essence, showing how the core message remains powerful even in its simplest form.

Detailed Comments

Clarity: The story and its compacted versions maintain a clear message about the power of math in achieving happiness.

Dynamism: The use of mathematical principles adds a dynamic element to the narrative.

Efficiency: The compacted versions efficiently convey the core message without losing its essence.

Effectiveness: Each version effectively communicates the relationship between math and happiness.

Realism: The story is grounded in realistic applications of math in personal well-being.

Series of Opinions

Math can be a powerful tool for achieving happiness.

Mathematical principles can simplify complex tasks.

Compacting stories can reveal their core essence.

Math can guide practical ways to achieve contentment.

Practical math can inspire a balanced lifestyle.

Sonnets

Sonnet 1: The Happiness Equation

In Harmonyville, where peace and joy reside, A mathematician sought a way to guide, Her life to lasting happiness and cheer, With math and logic, she would persevere.

She crafted formulas to light the way, To balance work and health and love each day. The Happiness Equation was her key, To finding joy and living happily.

Her town embraced her method, clear and bright, And found their lives were filled with pure delight. With math and wisdom, they did find their way, To happiness that lasts both night and day.

Her story proves that numbers hold the key, To living life with joy and harmony.

Sonnet 2: Emma's Mathematical Quest

In Harmonyville, Emma's quest began, To find a way to help her fellow man. With math and data, she did seek to find, A method to bring peace to troubled mind.

She gathered experts, analysts, and more, To build a model, solid to the core. The Happiness Equation was their aim, A formula to bring contentment's flame.

With careful thought and rigorous design, They crafted tools to help the heart align. The town embraced her method, clear and bright, And found their lives were filled with pure delight.

Her tale reminds us all of math's great power, To turn the tide and bring a brighter hour.

Epic Poems

Epic Poem 1: The Quest for Happiness

In Harmonyville, a town so bright and fair, A mathematician sought to ease despair. Emma was her name, a mind so keen, She sought to find a way to intervene.

With math and logic, she did seek to find, A method to bring peace to troubled mind. She crafted formulas to light the way, To balance work and health and love each day.

The Happiness Equation was her key, To finding joy and living happily. Her town embraced her method, clear and bright, And found their lives were filled with pure delight.

Through trials and tests, they forged a path so bright, Their formula a beacon in the night. Emma's story spread, a legend …

***

The relationship of math to finding a new practical way to reduce stress of daily life:

Microfiction Story (299 words)

In the bustling city of Numeria, a mathematician named Leo was overwhelmed by the stress of daily life. He decided to use his mathematical skills to find a practical way to reduce stress. Leo began by analyzing his daily activities and identifying the sources of his stress.

He created a mathematical model to quantify stress levels, assigning values to different activities based on their impact on his well-being. Leo called it the Stress Reduction Equation: $S=(W+C+T)R$, where S represented stress, W represented work, C represented chores, T represented time spent on stressful activities, and R represented relaxation time.

Leo tested his equation by applying it to his own life. He balanced his time between work, chores, and relaxation, ensuring that he had enough time to unwind each day. He tracked his stress levels over several weeks and found that the equation accurately predicted his overall well-being.

Encouraged by his success, Leo shared his findings with his friends and colleagues. People began to use the Stress Reduction Equation to guide their daily lives, making adjustments to achieve a balanced and stress-free lifestyle.

The city of Numeria thrived, with residents experiencing increased happiness and productivity.

Leo's story proved that math could be a powerful tool in reducing stress. His innovative approach inspired others to embrace mathematical principles in their pursuit of a balanced life. Numeria became a beacon of well-being, demonstrating that a balanced life, guided by math, could lead to lasting happiness.

Compacting the Story

250 Words

In Numeria, mathematician Leo was overwhelmed by daily stress. He decided to use his mathematical skills to find a practical way to reduce stress. Leo analyzed his daily activities and identified the sources of his stress.

He created the Stress Reduction Equation: $S=(W+C+T)R$, where S represented stress, W represented work, C represented chores, T represented time spent on stressful activities, and R represented relaxation time.

Leo tested his equation by applying it to his own life. He balanced his time between work, chores, and relaxation, ensuring enough time to unwind each day. He tracked his stress levels over several weeks and found the equation accurately predicted his well-being.

Encouraged by his success, Leo shared his findings with friends and colleagues. People used the Stress Reduction Equation to guide their lives, achieving a balanced and stress-free lifestyle. Numeria thrived, with residents experiencing increased happiness and productivity.

Leo's story proved math could be a powerful tool in reducing stress. His approach inspired others to embrace mathematical principles in their pursuit of a balanced life. Numeria became a beacon of well-being, demonstrating that a balanced life, guided by math, could lead to lasting happiness.

200 Words

In Numeria, mathematician Leo was overwhelmed by daily stress. He decided to use his mathematical skills to find a practical way to reduce stress. Leo analyzed his daily activities and identified the sources of his stress.

He created the Stress Reduction Equation: $S=(W+C+T)R$, where S represented stress, W represented work, C represented chores, T represented time spent on stressful activities, and R represented relaxation time.

Leo tested his equation by applying it to his own life. He balanced his time between work, chores, and relaxation, ensuring enough time to unwind each day. He tracked his stress levels over several weeks and found the equation accurately predicted his well-being.

Encouraged by his success, Leo shared his findings with friends and colleagues. People used the Stress Reduction Equation to guide their lives, achieving a balanced and stress-free lifestyle. Numeria thrived, with residents experiencing increased happiness and productivity.

Leo's story proved math could be a powerful tool in reducing stress. His approach inspired others to embrace mathematical principles in their pursuit of a balanced life.

150 Words

In Numeria, mathematician Leo was overwhelmed by daily stress. He decided to use his mathematical skills to find a practical way to reduce stress. Leo analyzed his daily activities and identified the sources of his stress.

He created the Stress Reduction Equation: $S=(W+C+T)R$, where S represented stress, W represented work, C represented chores, T represented time spent on stressful activities, and R represented relaxation time.

Leo tested his equation by applying it to his own life. He balanced his time between work, chores, and relaxation, ensuring enough time to unwind each day. He tracked his stress levels and found the equation accurately predicted his well-being.

Encouraged by his success, Leo shared his findings with friends and colleagues. People used the Stress Reduction Equation to guide their lives, achieving a balanced and stress-free lifestyle. Numeria thrived, with residents experiencing increased happiness and productivity.

100 Words

In Numeria, mathematician Leo was overwhelmed by daily stress. He decided to use his mathematical skills to reduce stress. Leo analyzed his daily activities and identified the sources of his stress.

He created the Stress Reduction Equation: $S=(W+C+T)R$.

Leo tested his equation by applying it to his own life. He balanced his time between work, chores, and relaxation. He tracked his stress levels and found the equation accurately predicted his well-being.

Encouraged by his success, Leo shared his findings. Numeria thrived, with residents experiencing increased happiness and productivity.

50 Words

In Numeria, mathematician Leo used math to reduce stress. He created the Stress Reduction Equation: $S=(W+C+T)R$. Leo tested it in his life and shared his findings. Numeria thrived, with residents experiencing increased happiness and productivity.

25 Words

Mathematician Leo created the Stress Reduction Equation: $S=(W+C+T)R$. Numeria thrived, with residents experiencing increased happiness.

20 Words

Mathematician Leo created the Stress Reduction Equation. Numeria thrived, with residents experiencing increased happiness.

15 Words

Leo created the Stress Reduction Equation. Numeria thrived.

10 Words

Leo's equation reduced stress. Numeria thrived.

7 Words

Leo's equation reduced stress, increased happiness.

5 Words

Equation reduced stress, increased happiness.

4 Words

Equation reduced stress.

3 Words

Reduced stress.

2 Words

Stress reduced.

1 Word

Relaxed.

0 Words

(Silence)

Explanation

The story illustrates how math can be used to find a practical way to reduce stress in daily life. By compacting the story, we distill its essence, showing how the core message remains powerful even in its simplest form.

Detailed Comments

- **Clarity**: The story and its compacted versions maintain a clear message about the power of math in reducing stress.
- **Dynamism**: The use of mathematical principles adds a dynamic element to the narrative.
- **Efficiency**: The compacted versions efficiently convey the core message without losing its essence.
- **Effectiveness**: Each version effectively communicates the relationship between math and stress reduction.
- **Realism**: The story is grounded in realistic applications of math in personal well-being.

Series of Opinions

1. Math can be a powerful tool for reducing stress.
2. Mathematical principles can simplify complex tasks.
3. Compacting stories can reveal their core essence.
4. Math can guide practical ways to achieve relaxation.
5. Practical math can inspire a balanced lifestyle.

Sonnets

Sonnet 1: The Stress Reduction Equation

In Numeria, where life is fast and bright, A mathematician sought to ease the plight, Of daily stress that weighed upon his mind, With math and logic, he would seek to find.

He crafted formulas to light the way, To balance work and chores and rest each day. The Stress Reduction Equation was his key, To finding peace and living stress-free.

His town embraced his method, clear and bright, And found their lives were filled with pure delight. With math and wisdom, they did find their way, To happiness that lasts both night and day.

His story proves that numbers hold the key, To living life with joy and harmony.

Sonnet 2: Leo's Mathematical Quest

In Numeria, Leo's quest began, To find a way to help his fellow man. With math and data, he did seek to find, A method to bring peace to troubled mind.

He gathered experts, analysts, and more, To build a model, solid to the core. The Stress Reduction Equation was their aim, A formula to bring contentment's flame.

With careful thought and rigorous design, They crafted tools to help the heart align. The town embraced his method, clear and bright, And found their lives were filled with pure delight.

His tale reminds us all of math's great power, To turn the tide and bring a brighter hour.

Epic Poems

Epic Poem 1: The Quest for Peace

In Numeria, a city bright and grand, A mathematician took a daring stand. Leo was his name, a mind so keen, He sought to find a way to intervene.

The daily stress that weighed upon his mind, He sought to ease with formulas to find. He crafted tools to light the way, To balance work and chores and rest each day.

The Stress Reduction Equation was his key, To finding peace and living stress-free. His town embraced his method, clear and bright, And found their lives were filled with pure delight.

***

Highlights the relationship of math to finding a new practical greatly improve memory:

Microfiction Story (299 words)

In the bustling city of Numeria, a mathematician named Clara was determined to find a practical way to greatly improve memory. She believed that math could unlock the secrets of the mind. Clara began by analyzing patterns in memory retention and recall.

She created a mathematical model to quantify memory strength, assigning values to different activities based on their impact on memory. Clara called it the Memory Enhancement Equation: $M=(R+P+S)T$, where M represented memory strength, R represented repetition, P represented practice, S represented sleep, and T represented time.

Clara tested her equation by applying it to her own life. She balanced her time between repetition, practice, and sleep, ensuring that she had enough time to rest each day. She tracked her memory performance over several weeks and found that the equation accurately predicted her overall memory improvement.

Encouraged by her success, Clara shared her findings with her friends and colleagues. People began to use the Memory Enhancement Equation to guide their daily routines, making adjustments to achieve better memory retention. The city of Numeria thrived, with residents experiencing improved cognitive abilities and productivity.

Clara's story proved that math could be a powerful tool in enhancing memory. Her innovative approach inspired others to embrace mathematical principles in their pursuit of cognitive improvement. Numeria became a beacon of intellectual growth, demonstrating that a balanced life, guided by math, could lead to lasting mental sharpness.

Compacting the Story

250 Words

In Numeria, mathematician Clara sought a practical way to improve memory. She believed math could unlock the secrets of the mind. Clara analyzed patterns in memory retention and recall.

She created the Memory Enhancement Equation: $M=(R+P+S)T$, where M represented memory strength, R represented repetition, P represented practice, S represented sleep, and T represented time.

Clara tested her equation by applying it to her own life. She balanced her time between repetition, practice, and sleep, ensuring enough rest each day. She tracked her memory performance over several weeks and found the equation accurately predicted her memory improvement.

Encouraged by her success, Clara shared her findings with friends and colleagues. People used the Memory Enhancement Equation to guide their routines, achieving better memory retention. Numeria thrived, with residents experiencing improved cognitive abilities and productivity.

Clara's story proved math could be a powerful tool in enhancing memory. Her approach inspired others to embrace mathematical principles in their pursuit of cognitive improvement. Numeria became a beacon of intellectual growth, demonstrating that a balanced life, guided by math, could lead to lasting mental sharpness.

200 Words

In Numeria, mathematician Clara sought a practical way to improve memory. She believed math could unlock the secrets of the mind. Clara analyzed patterns in memory retention and recall.

She created the Memory Enhancement Equation: $M=(R+P+S)T$, where M represented memory strength, R represented repetition, P represented practice, S represented sleep, and T represented time.

Clara tested her equation by applying it to her own life. She balanced her time between repetition, practice, and sleep, ensuring enough rest each day. She tracked her memory performance over several weeks and found the equation accurately predicted her memory improvement.

Encouraged by her success, Clara shared her findings with friends and colleagues. People used the Memory Enhancement Equation to guide their routines, achieving better memory retention. Numeria thrived, with residents experiencing improved cognitive abilities and productivity.

Clara's story proved math could be a powerful tool in enhancing memory. Her approach inspired others to embrace mathematical principles in their pursuit of cognitive improvement.

150 Words

In Numeria, mathematician Clara sought a practical way to improve memory. She believed math could unlock the secrets of the mind. Clara analyzed patterns in memory retention and recall.

She created the Memory Enhancement Equation: $M=(R+P+S)T$, where M represented memory strength, R represented repetition, P represented practice, S represented sleep, and T represented time.

Clara tested her equation by applying it to her own life. She balanced her time between repetition, practice, and sleep, ensuring enough rest each day. She tracked her memory performance and found the equation accurately predicted her memory improvement.

Encouraged by her success, Clara shared her findings with friends and colleagues. People used the Memory Enhancement Equation to guide their routines, achieving better memory retention. Numeria thrived, with residents experiencing improved cognitive abilities and productivity.

100 Words

In Numeria, mathematician Clara sought a practical way to improve memory. She believed math could unlock the secrets of the mind. Clara analyzed patterns in memory retention and recall.

She created the Memory Enhancement Equation: $M=(R+P+S)T$.

Clara tested her equation by applying it to her own life. She balanced her time between repetition, practice, and sleep. She tracked her memory performance and found the equation accurately predicted her memory improvement.

Encouraged by her success, Clara shared her findings. Numeria thrived, with residents experiencing improved cognitive abilities and productivity.

50 Words

In Numeria, mathematician Clara used math to improve memory. She created the Memory Enhancement Equation: M=(R+P+S)T. Clara tested it in her life and shared her findings. Numeria thrived, with residents experiencing improved cognitive abilities and productivity.

25 Words

Mathematician Clara created the Memory Enhancement Equation: M=(R+P+S)T. Numeria thrived, with residents experiencing improved cognitive abilities.

20 Words

Mathematician Clara created the Memory Enhancement Equation. Numeria thrived, with residents experiencing improved cognitive abilities.

15 Words

Clara created the Memory Enhancement Equation. Numeria thrived.

10 Words

Clara's equation improved memory. Numeria thrived.

7 Words

Clara's equation improved memory, productivity.

5 Words

Equation improved memory, productivity.

4 Words

Equation improved memory.

3 Words

Improved memory.

2 Words

Memory improved.

1 Word

Memory.

0 Words

(Silence)

Explanation

The story illustrates how math can be used to find a practical way to greatly improve memory. By compacting the story, we distill its essence, showing how the core message remains powerful even in its simplest form.

Detailed Comments

- **Clarity**: The story and its compacted versions maintain a clear message about the power of math in improving memory.

- **Dynamism**: The use of mathematical principles adds a dynamic element to the narrative.

- **Efficiency**: The compacted versions efficiently convey the core message without losing its essence.

- **Effectiveness**: Each version effectively communicates the relationship between math and memory improvement.

- **Realism**: The story is grounded in realistic applications of math in cognitive enhancement.

Series of Opinions

1. Math can be a powerful tool for improving memory.
2. Mathematical principles can simplify complex tasks.
3. Compacting stories can reveal their core essence.
4. Math can guide practical ways to achieve cognitive enhancement.
5. Practical math can inspire intellectual growth.

Sonnets

Sonnet 1: The Memory Enhancement Equation

In Numeria, where minds are sharp and bright, A mathematician sought to ease the plight, Of memory that fades with passing time, With math and logic, she would seek to climb.

She crafted formulas to light the way, To balance work and rest and practice each day. The Memory Enhancement Equation was her key, To finding strength and living mentally free.

Her town embraced her method, clear and bright, And found their minds were filled with pure delight. With math and wisdom, they did find their way, To memory that lasts both night and day.

Her story proves that numbers hold the key, To living life with mental clarity.

Sonnet 2: Clara's Mathematical Quest

In Numeria, Clara's quest began, To find a way to help her fellow man. With math and data, she did seek to find, A method to bring peace to troubled mind.

She gathered experts, analysts, and more, To build a model, solid to the core. The Memory Enhancement Equation was their aim, A formula to bring cognitive flame.

With careful thought and rigorous design, They crafted tools to help the mind align. The town embraced her method, clear and bright, And found their minds were filled with pure delight.

Her tale reminds us all of math's great power, To turn the tide and bring a brighter hour.

Epic Poem: The Quest for Memory

In Numeria, a city bright and grand, A mathematician took a daring stand. Clara was her name, a mind so keen, She sought to find a way to intervene.

The fading memory that plagued her mind, She sought to ease with formulas to find. She crafted tools to light the way, To balance work and rest and practice each day.

The Memory Enhancement Equation was her key, To finding strength and living mentally free. Her town embraced her method, clear and bright, And found their minds were filled with pure delight.

Through trials and tests, they forged a path so bright, Their formula a beacon in the night. Clara's story spread, a legend far and wide, Of how math's power turned the mental tide.

CHAPTER 24

Highlights the relationship of math to finding an excellent husband:

Microfiction Story (299 words)

In the vibrant city of Numeria, a mathematician named Emma was determined to find an excellent husband. She believed that math could help her make the best choice. Emma began by creating a list of qualities she desired in a partner, such as kindness, intelligence, and shared interests.

She assigned numerical values to each quality and developed a mathematical model to evaluate potential partners. Emma called it the Husband Selection Equation: $H=(K+I+S)C$, where H represented the husband score, K represented kindness, I represented intelligence, S represented shared interests, and C represented compatibility.

Emma tested her equation by applying it to her dating experiences. She went on dates and carefully evaluated each potential partner using her model. She tracked her satisfaction levels over several months and found that the equation accurately predicted her overall happiness with each partner.

Encouraged by her success, Emma shared her findings with her friends. They began to use the Husband Selection Equation to guide their own dating lives, making adjustments to find partners who met their criteria. The city of Numeria thrived, with residents experiencing increased happiness and fulfilling relationships.

Emma's story proved that math could be a powerful tool in finding an excellent husband. Her innovative approach inspired others to embrace mathematical principles in their pursuit of love. Numeria became a beacon of happy couples, demonstrating that a balanced life, guided by math, could lead to lasting joy.

Compacting the Story

250 Words

In Numeria, mathematician Emma sought to find an excellent husband. She believed math could help her make the best choice. Emma created a list of desired qualities, such as kindness, intelligence, and shared interests.

She assigned numerical values to each quality and developed the Husband Selection Equation: $H=(K+I+S)C$, where H represented the husband score, K represented kindness, I represented intelligence, S represented shared interests, and C represented compatibility.

Emma tested her equation by applying it to her dating experiences. She went on dates and evaluated each potential partner using her model. She tracked her satisfaction levels over several months and found the equation accurately predicted her happiness with each partner.

Encouraged by her success, Emma shared her findings with friends. They used the Husband Selection Equation to guide their own dating lives, finding partners who met their criteria. Numeria thrived, with residents experiencing increased happiness and fulfilling relationships.

Emma's story proved math could be a powerful tool in finding an excellent husband. Her approach inspired others to embrace mathematical principles in their pursuit of love. Numeria became a beacon of happy couples, demonstrating that a balanced life, guided by math, could lead to lasting joy.

200 Words

In Numeria, mathematician Emma sought to find an excellent husband. She believed math could help her make the best choice. Emma created a list of desired qualities, such as kindness, intelligence, and shared interests.

She assigned numerical values to each quality and developed the Husband Selection Equation: $H=(K+I+S)C$, where H represented the husband score, K represented kindness, I represented intelligence, S represented shared interests, and C represented compatibility.

Emma tested her equation by applying it to her dating experiences. She went on dates and evaluated each potential partner using her model. She tracked her satisfaction levels over several months and found the equation accurately predicted her happiness with each partner.

Encouraged by her success, Emma shared her findings with friends. They used the Husband Selection Equation to guide their own dating lives, finding partners who met their criteria. Numeria thrived, with residents experiencing increased happiness and fulfilling relationships.

Emma's story proved math could be a powerful tool in finding an excellent husband. Her approach inspired others to embrace mathematical principles in their pursuit of love.

150 Words

In Numeria, mathematician Emma sought to find an excellent husband. She believed math could help her make the best choice. Emma created a list of desired qualities, such as kindness, intelligence, and shared interests.

She assigned numerical values to each quality and developed the Husband Selection Equation: $H=(K+I+S)C$, where H represented the husband score, K represented kindness, I represented intelligence, S represented shared interests, and C represented compatibility.

Emma tested her equation by applying it to her dating experiences. She went on dates and evaluated each potential partner using her model. She tracked her satisfaction levels and found the equation accurately predicted her happiness with each partner.

Encouraged by her success, Emma shared her findings. Numeria thrived, with residents experiencing increased happiness and fulfilling relationships.

Emma's story proved math could be a powerful tool in finding an excellent husband.

100 Words

In Numeria, mathematician Emma sought to find an excellent husband. She believed math could help her make the best choice. Emma created a list of desired qualities, such as kindness, intelligence, and shared interests.

She assigned numerical values to each quality and developed the Husband Selection Equation: $H=(K+I+S)C$.

Emma tested her equation by applying it to her dating experiences. She tracked her satisfaction levels and found the equation accurately predicted her happiness with each partner.

Encouraged by her success, Emma shared her findings. Numeria thrived, with residents experiencing increased happiness.

50 Words

In Numeria, mathematician Emma used math to find an excellent husband. She created the Husband Selection Equation: $H=(K+I+S)C$. Emma tested it in her dating experiences and shared her findings. Numeria thrived, with residents experiencing increased happiness.

25 Words

Mathematician Emma created the Husband Selection Equation: $H=(K+I+S)C$. Numeria thrived, with residents experiencing increased happiness.

20 Words

Mathematician Emma created the Husband Selection Equation. Numeria thrived, with residents experiencing increased happiness.

15 Words

Emma created the Husband Selection Equation. Numeria thrived.

10 Words

Emma's equation found husbands. Numeria thrived.

7 Words

Emma's equation found husbands, increased happiness.

5 Words

Equation found husbands, increased happiness.

4 Words

Equation found husbands.

3 Words

Found husbands.

2 Words

Husbands found.

1 Word

Husbands.

0 Words

(Silence)

Explanation

The story illustrates how math can be used to find an excellent husband. By compacting the story, we distill its essence, showing how the core message remains powerful even in its simplest form.

Detailed Comments

- **Clarity**: The story and its compacted versions maintain a clear message about the power of math in finding a partner.
- **Dynamism**: The use of mathematical principles adds a dynamic element to the narrative.
- **Efficiency**: The compacted versions efficiently convey the core message without losing its essence.
- **Effectiveness**: Each version effectively communicates the relationship between math and partner selection.
- **Realism**: The story is grounded in realistic applications of math in personal relationships.

Series of Opinions

1. Math can be a powerful tool for finding a partner.
2. Mathematical principles can simplify complex tasks.
3. Compacting stories can reveal their core essence.
4. Math can guide practical ways to achieve fulfilling relationships.
5. Practical math can inspire happiness in personal life.

Sonnets

Sonnet 1: The Husband Selection Equation

In Numeria, where love and logic blend, A mathematician sought a way to mend, Her heart with numbers, formulas, and more, To find a husband she could not ignore.

She crafted equations to light the way, To balance kindness, smarts, and love each day. The Husband Selection Equation was her key, To finding love and living happily.

Her town embraced her method, clear and bright, And found their hearts were filled with pure delight. With math and wisdom, they did find their way, To happiness that lasts both night and day.

Her story proves that numbers hold the key, To living life with joy and harmony.

Sonnet 2: Emma's Mathematical Quest

In Numeria, Emma's quest began, To find a way to help her fellow man. With math and data, she did seek to find, A method to bring peace to troubled mind.

She gathered experts, analysts, and more, To build a model, solid to the core. The Husband Selection Equation was their aim, A formula to bring love's gentle flame.

With careful thought and rigorous design, They crafted tools to help the heart align. The town embraced her method, clear and bright, And found their hearts were filled with pure delight.

Her tale reminds us all of math's great power, To turn the tide and bring a brighter hour.

Epic Poems

Epic Poem 1: The Quest for Love

In Numeria, a city bright and grand, A mathematician took a daring stand. Emma was her name, a mind so keen, She sought to find a way to intervene.

The search for love that plagued her heart and mind, She sought to ease with formulas to find. She crafted tools to light the way, To balance kindness, smarts, and love each day.

The Husband Selection Equation was her key, To finding love and living happily. Her town embraced her method, clear and bright, And found their hearts were filled with pure delight.

Through trials and tests, they forged a path so bright, Their formula a …

***

Another example:

Leo's equation won prizes, jackpot.

5 Words

Equation won prizes, jackpot.

4 Words

Equation won jackpot.

3 Words

Won jackpot.

2 Words

Jackpot won.

1 Word

Jackpot.

0 Words

(Silence)

Explanation

The story illustrates how math can be used to find a practical way to pick a winning lottery ticket. By compacting the story, we distill its essence, showing how the core message remains powerful even in its simplest form.

Detailed Comments

- **Clarity**: The story and its compacted versions maintain a clear message about the power of math in predicting lottery outcomes.

- **Dynamism**: The use of mathematical principles adds a dynamic element to the narrative.

- **Efficiency**: The compacted versions efficiently convey the core message without losing its essence.

- **Effectiveness**: Each version effectively communicates the relationship between math and lottery prediction.

- **Realism**: The story is grounded in realistic applications of math in predicting outcomes.

Series of Opinions

1. Math can be a powerful tool for predicting outcomes.
2. Mathematical principles can simplify complex tasks.
3. Compacting stories can reveal their core essence.
4. Math can guide practical ways to achieve success.
5. Practical math can inspire confidence in decision-making.

Sonnets

Sonnet 1: The Lottery Prediction Equation

In Numeria, where luck and logic blend, A mathematician sought a way to mend, The odds of winning with a formula bright, To pick the numbers that would bring delight.

He crafted equations to light the way, To balance patterns
…

***

Highlights the relationship of math to finding an ideal wife:

Microfiction Story (299 words)

In the bustling city of Numeria, a mathematician named Leo was determined to find his ideal wife. He believed that math could help him make the best choice. Leo began by creating a list of qualities he desired in a partner, such as kindness, intelligence, and shared interests.

He assigned numerical values to each quality and developed a mathematical model to evaluate potential partners. Leo called it the Wife Selection Equation: $W=(K+I+S)C$, where W represented the wife score, K represented kindness, I represented intelligence, S represented shared interests, and C represented compatibility.

Leo tested his equation by applying it to his dating experiences. He went on dates and carefully evaluated each potential partner using his model. He tracked his satisfaction levels over several months and found that the equation accurately predicted his overall happiness with each partner.

Encouraged by his success, Leo shared his findings with his friends. They began to use the Wife Selection Equation to guide their own dating lives, making adjustments to find partners who met their criteria. The city of Numeria thrived, with residents experiencing increased happiness and fulfilling relationships.

Leo's story proved that math could be a powerful tool in finding an ideal wife. His innovative approach inspired others to embrace mathematical principles in their pursuit of love. Numeria became a beacon of happy couples, demonstrating that a balanced life, guided by math, could lead to lasting joy.

Compacting the Story

250 Words

In Numeria, mathematician Leo sought to find his ideal wife. He believed math could help him make the best choice. Leo created a list of desired qualities, such as kindness, intelligence, and shared interests.

He assigned numerical values to each quality and developed the Wife Selection Equation: $W=(K+I+S)C$, where W represented the wife score, K represented kindness, I represented intelligence, S represented shared interests, and C represented compatibility.

Leo tested his equation by applying it to his dating experiences. He went on dates and evaluated each potential partner using his model. He tracked his satisfaction levels over several months and found the equation accurately predicted his happiness with each partner.

Encouraged by his success, Leo shared his findings with friends. They used the Wife Selection Equation to guide their own dating lives, finding partners who met their criteria. Numeria thrived, with residents experiencing increased happiness and fulfilling relationships.

Leo's story proved math could be a powerful tool in finding an ideal wife. His approach inspired others to embrace mathematical principles in their pursuit of love. Numeria became a beacon of happy couples, demonstrating that a balanced life, guided by math, could lead to lasting joy.

200 Words

In Numeria, mathematician Leo sought to find his ideal wife. He believed math could help him make the best choice. Leo created a list of desired qualities, such as kindness, intelligence, and shared interests.

He assigned numerical values to each quality and developed the Wife Selection Equation: $W=(K+I+S)C$, where W represented the wife score, K represented kindness, I represented intelligence, S represented shared interests, and C represented compatibility.

Leo tested his equation by applying it to his dating experiences. He went on dates and evaluated each potential partner using his model. He tracked his satisfaction levels over several months and found the equation accurately predicted his happiness with each partner.

Encouraged by his success, Leo shared his findings with friends. They used the Wife Selection Equation to guide their own dating lives, finding partners who met their criteria. Numeria thrived, with residents experiencing increased happiness and fulfilling relationships.

Leo's story proved math could be a powerful tool in finding an ideal wife. His approach inspired others to embrace mathematical principles in their pursuit of love.

150 Words

In Numeria, mathematician Leo sought to find his ideal wife. He believed math could help him make the best choice. Leo created a list of desired qualities, such as kindness, intelligence, and shared interests.

He assigned numerical values to each quality and developed the Wife Selection Equation: $W=(K+I+S)C$, where W represented the wife score, K represented kindness, I represented intelligence, S represented shared interests, and C represented compatibility.

Leo tested his equation by applying it to his dating experiences. He went on dates and evaluated each potential partner using his model. He tracked his satisfaction levels and found the equation accurately predicted his happiness with each partner.

Encouraged by his success, Leo shared his findings. Numeria thrived, with residents experiencing increased happiness and fulfilling relationships.

Leo's story proved math could be a powerful tool in finding an ideal wife.

100 Words

In Numeria, mathematician Leo sought to find his ideal wife. He believed math could help him make the best choice. Leo created a list of desired qualities, such as kindness, intelligence, and shared interests.

He assigned numerical values to each quality and developed the Wife Selection Equation: $W=(K+I+S)C$.

Leo tested his equation by applying it to his dating experiences. He tracked his satisfaction levels and found the equation accurately predicted his happiness with each partner.

Encouraged by his success, Leo shared his findings. Numeria thrived, with residents experiencing increased happiness.

50 Words

In Numeria, mathematician Leo used math to find his ideal wife. He created the Wife Selection Equation: $W=(K+I+S)C$. Leo tested it in his dating experiences and shared his findings. Numeria thrived, with residents experiencing increased happiness.

25 Words

Mathematician Leo created the Wife Selection Equation: $W=(K+I+S)C$. Numeria thrived, with residents experiencing increased happiness.

20 Words

Mathematician Leo created the Wife Selection Equation. Numeria thrived, with residents experiencing increased happiness.

15 Words

Leo created the Wife Selection Equation. Numeria thrived.

10 Words

Leo's equation found wives. Numeria thrived.

7 Words

Leo's equation found wives, increased happiness.

5 Words

Equation found wives, increased happiness.

4 Words

Equation found wives.

3 Words

Found wives.

2 Words

Wives found.

1 Word

Wives.

0 Words

(Silence)

Explanation

The story illustrates how math can be used to find an ideal wife. By compacting the story, we distill its essence, showing how the core message remains powerful even in its simplest form.

Detailed Comments

- **Clarity**: The story and its compacted versions maintain a clear message about the power of math in finding a partner.
- **Dynamism**: The use of mathematical principles adds a dynamic element to the narrative.
- **Efficiency**: The compacted versions efficiently convey the core message without losing its essence.
- **Effectiveness**: Each version effectively communicates the relationship between math and partner selection.
- **Realism**: The story is grounded in realistic applications of math in personal relationships.

Series of Opinions

1. Math can be a powerful tool for finding a partner.
2. Mathematical principles can simplify complex tasks.
3. Compacting stories can reveal their core essence.
4. Math can guide practical ways to achieve fulfilling relationships.
5. Practical math can inspire happiness in personal life.

Sonnets

Sonnet 1: The Wife Selection Equation

In Numeria, where love and logic blend, A mathematician sought a way to mend, His heart with numbers, formulas, and more, To find a wife he could not ignore.

He crafted equations to light the way, To balance kindness, smarts, and love each day. The Wife Selection Equation was his key, To finding love and living happily.

His town embraced his method, clear and bright, And found their hearts were filled with pure delight. With math and wisdom, they did find their way, To happiness that lasts both night and day.

His story proves that numbers hold the key, To living life with joy and harmony.

Sonnet 2: Leo's Mathematical Quest

In Numeria, Leo's quest began, To find a way to help his fellow man. With math and data, he did seek to find, A method to bring peace to troubled mind.

He gathered experts, analysts, and more, To build a model, solid to the core. The Wife Selection Equation was their aim, A formula to bring love's gentle flame.

With careful thought and rigorous design, They crafted tools to help the heart align. The town embraced his method, clear and bright, And found their hearts were filled with pure delight.

His tale reminds us all of math's great power, To turn the tide and bring a brighter hour.

Epic Poem: The Quest for Love

In Numeria, a city bright and grand, A mathematician took a daring stand. Leo was his name, a mind so keen, He sought to find a way to intervene.

The search for love that plagued his heart and mind, He sought to ease with formulas to find. He crafted tools to light the way, To balance kindness, smarts, and love each day.

The Wife Selection Equation was his key, To finding love and living happily. His town embraced his method, clear and bright, And found their hearts were filled with pure delight.

Through trials and tests, they forged a path so bright, Their formula a beacon in …

CHAPTER 25

Highlights the relationship of math to finding a new practical to find the best community to live in:

Microfiction Story (299 words)

In the bustling city of Numeria, a mathematician named Clara was determined to find the best community to live in. She believed that math could help her make the best choice. Clara began by creating a list of factors that were important to her, such as safety, education, healthcare, and community engagement.

She assigned numerical values to each factor and developed a mathematical model to evaluate different communities. Clara called it the Community Selection Equation: $C=(S+E+H+G)P$, where C represented the community score, S represented safety, E represented education, H represented healthcare, G represented community engagement, and P represented population density.

Clara tested her equation by applying it to various communities. She collected data on each factor and calculated the community scores. She visited the top-scoring communities to see if they matched her expectations. Clara found that the equation accurately predicted the quality of life in each community.

Encouraged by her success, Clara shared her findings with her friends and colleagues. People began to use the Community Selection Equation to guide their decisions on where to live, making adjustments to find communities that met their criteria. The city of Numeria thrived, with residents experiencing increased happiness and a stronger sense of community.

Clara's story proved that math could be a powerful tool in finding the best community to live in. Her innovative approach inspired others to embrace mathematical principles in their pursuit of a better life. Numeria became a beacon of thriving communities, demonstrating that a balanced life, guided by math, could lead to lasting happiness.

Compacting the Story

250 Words

In Numeria, mathematician Clara sought to find the best community to live in. She believed math could help her make the best choice. Clara created a list of important factors, such as safety, education, healthcare, and community engagement.

She assigned numerical values to each factor and developed the Community Selection Equation: $C=(S+E+H+G)P$, where C represented the community score, S represented safety, E represented education, H represented healthcare, G represented community engagement, and P represented population density.

Clara tested her equation by applying it to various communities. She collected data on each factor and calculated the community scores. She visited the top-scoring communities to see if they matched her expectations. Clara found that the equation accurately predicted the quality of life in each community.

Encouraged by her success, Clara shared her findings with friends and colleagues. People used the Community Selection Equation to guide their decisions on where to live, finding communities that met their criteria. Numeria thrived, with residents experiencing increased happiness and a stronger sense of community.

Clara's story proved math could be a powerful tool in finding the best community to live in. Her approach inspired others to embrace mathematical principles in their pursuit of a better life. Numeria became a beacon of thriving communities, demonstrating that a balanced life, guided by math, could lead to lasting happiness.

200 Words

In Numeria, mathematician Clara sought to find the best community to live in. She believed math could help her make the best choice. Clara created a list of important factors, such as safety, education, healthcare, and community engagement.

She assigned numerical values to each factor and developed the Community Selection Equation: $C=(S+E+H+G)P$, where C represented the community score, S represented safety, E represented education, H represented healthcare, G represented community engagement, and P represented population density.

Clara tested her equation by applying it to various communities. She collected data on each factor and calculated the community scores. She visited the top-scoring communities to see if they matched her expectations. Clara found that the equation accurately predicted the quality of life in each community.

Encouraged by her success, Clara shared her findings with friends and colleagues. People used the Community Selection Equation to guide their decisions on where to live, finding communities that met their criteria. Numeria thrived, with residents experiencing increased happiness and a stronger sense of community.

Clara's story proved math could be a powerful tool in finding the best community to live in.

150 Words

In Numeria, mathematician Clara sought to find the best community to live in. She believed math could help her make the best choice. Clara created a list of important factors, such as safety, education, healthcare, and community engagement.

She assigned numerical values to each factor and developed the Community Selection Equation: $C=(S+E+H+G)P$, where C represented the community score, S represented safety, E represented education, H represented healthcare, G represented community engagement, and P represented population density.

Clara tested her equation by applying it to various communities. She collected data on each factor and calculated the community scores. She visited the top-scoring communities to see if they matched her expectations. Clara found that the equation accurately predicted the quality of life in each community.

Encouraged by her success, Clara shared her findings. Numeria thrived, with residents experiencing increased happiness and a stronger sense of community.

100 Words

In Numeria, mathematician Clara sought to find the best community to live in. She believed math could help her make the best choice. Clara created a list of important factors, such as safety, education, healthcare, and community engagement.

She assigned numerical values to each factor and developed the Community Selection Equation: $C=(S+E+H+G)P$.

Clara tested her equation by applying it to various communities. She collected data on each factor and calculated the community scores. Clara found that the equation accurately predicted the quality of life in each community.

Encouraged by her success, Clara shared her findings.

50 Words

In Numeria, mathematician Clara used math to find the best community to live in. She created the Community Selection

Equation: $C=(S+E+H+G)P$. Clara tested it and shared her findings. Numeria thrived, with residents experiencing increased happiness.

25 Words

Mathematician Clara created the Community Selection Equation: $C=(S+E+H+G)P$. Numeria thrived, with residents experiencing increased happiness.

20 Words

Mathematician Clara created the Community Selection Equation. Numeria thrived, with residents experiencing increased happiness.

15 Words

Clara created the Community Selection Equation. Numeria thrived.

10 Words

Clara's equation found communities. Numeria thrived.

7 Words

Clara's equation found communities, increased happiness.

5 Words

Equation found communities, increased happiness.

4 Words

Equation found communities.

3 Words

Found communities.

2 Words

Communities found.

1 Word

Communities.

0 Words

(Silence)

Explanation

The story illustrates how math can be used to find the best community to live in. By compacting the story, we distill its essence, showing how the core message remains powerful even in its simplest form.

Detailed Comments

- **Clarity**: The story and its compacted versions maintain a clear message about the power of math in finding a community.

- **Dynamism**: The use of mathematical principles adds a dynamic element to the narrative.
- **Efficiency**: The compacted versions efficiently convey the core message without losing its essence.
- **Effectiveness**: Each version effectively communicates the relationship between math and community selection.
- **Realism**: The story is grounded in realistic applications of math in personal decisions.

Series of Opinions

1. Math can be a powerful tool for finding a community.
2. Mathematical principles can simplify complex tasks.
3. Compacting stories can reveal their core essence.
4. Math can guide practical ways to achieve fulfilling living conditions.
5. Practical math can inspire happiness in personal life.

Sonnets

Sonnet 1: The Community Selection Equation

In Numeria, where logic and love blend, A mathematician sought a way to mend, Her life with numbers, formulas, and more, To find a community she could adore.

She crafted equations to light the way, To balance safety, health, and love each day. The Community Selection Equation was her key, To finding peace and living happily.

Her town embraced her method, clear and bright, And found their lives were filled with pure delight. With math and wisdom, they did find their way, To happiness that lasts both night and day.

Her story proves that numbers hold the key, To living life with joy and harmony.

Sonnet 2: Clara's Mathematical Quest

In Numeria, Clara's quest began, To find a way to help her fellow man. With math and data, she did seek to find, A method to bring peace to troubled mind.

She gathered experts, analysts, and more, To build a model, solid to the core. The Community Selection Equation was their aim, A formula to bring love's gentle flame.

With careful thought and rigorous design, They crafted tools to help the heart align. The town embraced her method, clear and bright, And found their lives were filled with pure delight.

Her tale reminds us all of math's great power, To turn the tide and bring a brighter hour.

***

Highlights the relationship of math for a woman to find a new fashion style best suited for her:

Microfiction Story (299 words)

Inthe vibrant city of Numeria, a woman named Clara was determined to find a new fashion style best suited for her. She believed that math could help her make the best choice. Clara began by creating a list of factors that were important to her, such as comfort, color, fit, and trendiness.

She assigned numerical values to each factor and developed a mathematical model to evaluate different fashion styles. Clara called it the Fashion Selection Equation: $F=(C+Co+F+T)P$, where F represented the fashion score, C represented comfort, Co represented color, F represented fit, T represented trendiness, and P represented personal preference.

Clara tested her equation by applying it to various outfits. She tried on different styles and carefully evaluated each one using her model. She tracked her satisfaction levels over several weeks and found that the equation accurately predicted her overall happiness with each outfit.

Encouraged by her success, Clara shared her findings with her friends. They began to use the Fashion Selection Equation to guide their own fashion choices, making adjustments to find styles that suited them best. The city of Numeria thrived, with residents experiencing increased confidence and a stronger sense of personal style.

Clara's story proved that math could be a powerful tool in finding the best fashion style. Her innovative approach inspired others to embrace mathematical principles in their pursuit of personal expression. Numeria became a beacon of stylish individuals, demonstrating that a balanced life, guided by math, could lead to lasting confidence and happiness.

Compacting the Story

250 Words

In Numeria, Clara sought to find a new fashion style best suited for her. She believed math could help her make the best choice. Clara created a list of important factors, such as comfort, color, fit, and trendiness.

She assigned numerical values to each factor and developed the Fashion Selection Equation: $F=(C+Co+F+T)P$, where F represented the fashion score, C represented comfort, Co represented color, F represented fit, T represented trendiness, and P represented personal preference.

Clara tested her equation by applying it to various outfits. She tried on different styles and evaluated each one using her model. She tracked her satisfaction levels over several weeks and found the equation accurately predicted her happiness with each outfit.

Encouraged by her success, Clara shared her findings with friends. They used the Fashion Selection Equation to guide their own fashion choices, finding styles that suited them best. Numeria thrived, with residents experiencing increased confidence and a stronger sense of personal style.

Clara's story proved math could be a powerful tool in finding the best fashion style. Her approach inspired others to embrace mathematical principles in their pursuit of personal expression. Numeria became a beacon of stylish individuals, demonstrating that a balanced life, guided by math, could lead to lasting confidence and happiness.

200 Words

In Numeria, Clara sought to find a new fashion style best suited for her. She believed math could help her make the best choice. Clara created a list of important factors, such as comfort, color, fit, and trendiness.

She assigned numerical values to each factor and developed the Fashion Selection Equation: $F=(C+Co+F+T)P$, where F represented the fashion score, C represented comfort, Co represented color, F represented fit, T represented trendiness, and P represented personal preference.

Clara tested her equation by applying it to various outfits. She tried on different styles and evaluated each one using her model. She tracked her satisfaction levels over several weeks and found the equation accurately predicted her happiness with each outfit.

Encouraged by her success, Clara shared her findings with friends. They used the Fashion Selection Equation to guide their own fashion choices, finding styles that suited them best. Numeria thrived, with residents experiencing increased confidence and a stronger sense of personal style.

Clara's story proved math could be a powerful tool in finding the best fashion style. Her approach inspired others to embrace mathematical principles in their pursuit of personal expression.

150 Words

In Numeria, Clara sought to find a new fashion style best suited for her. She believed math could help her make the best choice. Clara created a list of important factors, such as comfort, color, fit, and trendiness.

She assigned numerical values to each factor and developed the Fashion Selection Equation: $F=(C+Co+F+T)P$, where F represented the fashion score, C represented comfort, Co represented color, F represented fit, T represented trendiness, and P represented personal preference.

Clara tested her equation by applying it to various outfits. She tried on different styles and evaluated each one using her model. She tracked her satisfaction levels and found the equation accurately predicted her happiness with each outfit.

Encouraged by her success, Clara shared her findings. Numeria thrived, with residents experiencing increased confidence and a stronger sense of personal style.

Clara's story proved math could be a powerful tool in finding the best fashion style.

100 Words

In Numeria, Clara sought to find a new fashion style best suited for her. She believed math could help her make the best choice. Clara created a list of important factors, such as comfort, color, fit, and trendiness.

She assigned numerical values to each factor and developed the Fashion Selection Equation: $F=(C+Co+F+T)P$.

Clara tested her equation by applying it to various outfits. She tracked her satisfaction levels and found the equation accurately predicted her happiness with each outfit.

Encouraged by her success, Clara shared her findings. Numeria thrived, with residents experiencing increased confidence.

50 Words

In Numeria, Clara used math to find the best fashion style. She created the Fashion Selection Equation: $F=(C+Co+F+T)P$. Clara tested it and shared her findings. Numeria thrived, with residents experiencing increased confidence.

25 Words

Clara created the Fashion Selection Equation: $F=(C+Co+F+T)P$. Numeria thrived, with residents experiencing increased confidence.

20 Words

Clara created the Fashion Selection Equation. Numeria thrived, with residents experiencing increased confidence.

15 Words

Clara created the Fashion Selection Equation. Numeria thrived.

10 Words

Clara's equation found fashion. Numeria thrived.

7 Words

Clara's equation found fashion, increased confidence.

5 Words

Equation found fashion, increased confidence.

4 Words

Equation found fashion.

3 Words

Found fashion.

2 Words

Fashion found.

1 Word

Fashion.

0 Words

(Silence)

Explanation

The story illustrates how math can be used to find the best fashion style. By compacting the story, we distill its essence, showing how the core message remains powerful even in its simplest form.

Detailed Comments

- **Clarity**: The story and its compacted versions maintain a clear message about the power of math in finding a fashion style.

- **Dynamism**: The use of mathematical principles adds a dynamic element to the narrative.

- **Efficiency**: The compacted versions efficiently convey the core message without losing its essence.

- **Effectiveness**: Each version effectively communicates the relationship between math and fashion selection.

- **Realism**: The story is grounded in realistic applications of math in personal decisions.

Series of Opinions

1. Math can be a powerful tool for finding a fashion style.
2. Mathematical principles can simplify complex tasks.
3. Compacting stories can reveal their core essence.
4. Math can guide practical ways to achieve personal expression.
5. Practical math can inspire confidence in personal style.

Sonnet: The Fashion Selection Equation

In Numeria, where logic and style blend, A mathematician sought a way to mend, Her wardrobe with numbers, formulas, and more, To find a fashion she could not ignore.

She crafted equations to light the way, To balance comfort, fit, and trend each day. The Fashion Selection Equation was her key, To finding style and living happily.

Her town embraced her method, clear and bright, And found their looks were filled with pure delight. With math and wisdom, they did find their way, To confidence that lasts both night and day.

Her story proves ...

***

Highlights the relationship of math to finding a new practical hobby best suited for a person:

Microfiction Story (299 words)

In the vibrant city of Numeria, a young woman named Clara was determined to find a new hobby that would bring her joy and fulfillment. She believed that math could help her make the best choice. Clara began by listing her interests and the qualities she desired in a hobby, such as creativity, relaxation, and social interaction.

She assigned numerical values to each quality and developed a mathematical model to evaluate different hobbies. Clara called it the Hobby Selection Equation: $H=(C+R+S)T$, where H represented the hobby score, C represented creativity, R represented relaxation, S represented social interaction, and T represented time commitment.

Clara tested her equation by applying it to various hobbies. She collected data on each activity and calculated the hobby scores. She tried out the top-scoring hobbies to see if they matched her expectations. Clara found that the equation accurately predicted her enjoyment and satisfaction with each hobby.

Encouraged by her success, Clara shared her findings with her friends and colleagues. People began to use the Hobby Selection Equation to guide their decisions on which hobbies to pursue, making adjustments to find activities that met their criteria. The city of Numeria thrived, with residents experiencing increased happiness and a stronger sense of community.

Clara's story proved that math could be a powerful tool in finding the best hobby. Her innovative approach inspired others to embrace mathematical principles in their pursuit of personal fulfillment. Numeria became a beacon of thriving hobbies, demonstrating that a balanced life, guided by math, could lead to lasting happiness.

Compacting the Story

250 Words

In Numeria, Clara sought to find a new hobby that would bring her joy and fulfillment. She believed math could help her make the best choice. Clara listed her interests and the qualities she desired in a hobby, such as creativity, relaxation, and social interaction.

She assigned numerical values to each quality and developed the Hobby Selection Equation: $H=(C+R+S)T$, where H represented the hobby score, C represented creativity, R represented relaxation, S represented social interaction, and T represented time commitment.

Clara tested her equation by applying it to various hobbies. She collected data on each activity and calculated the hobby scores. She tried out the top-scoring hobbies to see if they matched her expectations. Clara found that the equation accurately predicted her enjoyment and satisfaction with each hobby.

Encouraged by her success, Clara shared her findings with friends and colleagues. People used the Hobby Selection Equation to guide their decisions on which hobbies to pursue, finding activities that met their criteria. Numeria thrived, with residents experiencing increased happiness and a stronger sense of community.

Clara's story proved math could be a powerful tool in finding the best hobby. Her approach inspired others to embrace mathematical principles in their pursuit of personal fulfillment. Numeria became a beacon of thriving hobbies, demonstrating that a balanced life, guided by math, could lead to lasting happiness.

200 Words

In Numeria, Clara sought to find a new hobby that would bring her joy and fulfillment. She believed math could help her make the best choice. Clara listed her interests and the qualities she desired in a hobby, such as creativity, relaxation, and social interaction.

She assigned numerical values to each quality and developed the Hobby Selection Equation: $H=(C+R+S)T$, where H represented the hobby score, C represented creativity, R represented relaxation, S represented social interaction, and T represented time commitment.

Clara tested her equation by applying it to various hobbies. She collected data on each activity and calculated the hobby scores. She tried out the top-scoring hobbies to see if they matched her expectations. Clara found that the equation accurately predicted her enjoyment and satisfaction with each hobby.

Encouraged by her success, Clara shared her findings with friends and colleagues. People used the Hobby Selection Equation to guide their decisions on which hobbies to pursue, finding activities that met their criteria. Numeria thrived, with residents experiencing increased happiness and a stronger sense of community.

Clara's story proved math could be a powerful tool in finding the best hobby.

150 Words

In Numeria, Clara sought to find a new hobby that would bring her joy and fulfillment. She believed math could help her make the best choice. Clara listed her interests and the qualities she desired in a hobby, such as creativity, relaxation, and social interaction.

She assigned numerical values to each quality and developed the Hobby Selection Equation: $H=(C+R+S)T$, where H represented the hobby score, C represented creativity, R represented relaxation, S represented social interaction, and T represented time commitment.

Clara tested her equation by applying it to various hobbies. She collected data on each activity and calculated the hobby scores. She tried out the top-scoring hobbies to see if they matched her expectations. Clara found that the equation accurately predicted her enjoyment and satisfaction with each hobby.

Encouraged by her success, Clara shared her findings. Numeria thrived, with residents experiencing increased happiness and a stronger sense of community.

100 Words

In Numeria, Clara sought to find a new hobby that would bring her joy and fulfillment. She believed math could help her make the best choice. Clara listed her interests and the qualities she desired in a hobby, such as creativity, relaxation, and social interaction.

She assigned numerical values to each quality and developed the Hobby Selection Equation: $H=(C+R+S)T$.

Clara tested her equation by applying it to various hobbies. She collected data on each activity and calculated the hobby scores. Clara found that the equation accurately predicted her enjoyment and satisfaction with each hobby.

Encouraged by her success, Clara shared her findings.

50 Words

In Numeria, Clara used math to find a new hobby. She created the Hobby Selection Equation: $H=(C+R+S)T$. Clara tested it and shared her findings. Numeria thrived, with residents experiencing increased happiness.

25 Words

Clara created the Hobby Selection Equation: $H=(C+R+S)T$. Numeria thrived, with residents experiencing increased happiness.

20 Words

Clara created the Hobby Selection Equation. Numeria thrived, with residents experiencing increased happiness.

15 Words

Clara created the Hobby Selection Equation. Numeria thrived.

10 Words

Clara's equation found hobbies. Numeria thrived.

7 Words

Clara's equation found hobbies, increased happiness.

5 Words

Equation found hobbies, happiness.

4 Words

Equation found hobbies.

3 Words

Found hobbies.

2 Words

Hobbies found.

1 Word

Hobbies.

0 Words

(Silence)

Explanation

The story illustrates how math can be used to find a new practical hobby best suited for a person. By compacting the story, we distill its essence, showing how the core message remains powerful even in its simplest form.

Detailed Comments

- **Clarity**: The story and its compacted versions maintain a clear message about the power of math in finding a hobby.
- **Dynamism**: The use of mathematical principles adds a dynamic element to the narrative.
- **Efficiency**: The compacted versions efficiently convey the core message without losing its essence.
- **Effectiveness**: Each version effectively communicates the relationship between math and hobby selection.
- **Realism**: The story is grounded in realistic applications of math in personal decisions.

Series of Opinions

1. Math can be a powerful tool for finding a hobby.
2. Mathematical principles can simplify complex tasks.
3. Compacting stories can reveal their core essence.
4. Math can guide practical ways to achieve fulfilling activities.
5. Practical math can inspire happiness in personal life.

Sonnets

Sonnet 1: The Hobby Selection Equation

In Numeria, where logic and joy blend, A mathematician sought a way to mend, Her life with numbers, formulas, and more, To find a hobby she could not ignore.

She crafted equations to light the way, To balance creativity, rest, and play. The Hobby Selection Equation was her key, To finding peace and living happily.

Her town embraced her method, clear and bright, And found their lives were filled with pure delight. With math and wisdom, they did find their way, To happiness that lasts both night and day.

Her story proves that numbers hold the key, To living life with joy and harmony.

Sonnet 2: Clara's Mathematical Quest

In Numeria, Clara's quest began, To find a way to help her fellow man. With math and data, she did seek to find, A method to bring peace to troubled mind.

She gathered experts, analysts, and more, To build a model, solid to the core. The Hobby Selection Equation was their aim, A formula to bring joy's gentle flame.

With careful thought and rigorous design, They crafted tools to help the heart align. The town embraced her method, clear and bright, And found their lives were filled with pure delight.

Her tale reminds us all of math's great power, To turn the tide and bring a brighter hour.

Epic Poem: The Quest for Hobbies

In Numeria, a city bright and grand, A mathematician took a daring stand. Clara was her name, a mind so keen, She sought to find a way to intervene.

The search for hobbies …

***

Highlights the relationship of math to escape a bad relationship:

Microfiction Story (299 words)

In the bustling city of Numeria, a mathematician named Emma found herself trapped in a toxic relationship. She decided to use her mathematical skills to find a way out. Emma began by analyzing her relationship, assigning numerical values to different aspects such as happiness, stress, and support.

She created a mathematical model to evaluate the overall health of her relationship. Emma called it the Relationship Health Equation:

$$R = (H - S + Su$$

)

T

, where

R

represented the relationship score,

H

represented happiness,

S

represented stress,

S_u

represented support, and

T

represented time.

Emma tested her equation by applying it to her relationship. She tracked her happiness, stress, and support levels over several months and found that the equation accurately predicted the decline in her relationship's health. The relationship score was consistently low, confirming her need to leave.

Encouraged by her findings, Emma devised a plan to escape. She used her mathematical skills to create a budget, ensuring she had enough resources to support herself. She calculated the optimal time to leave, considering factors such as work commitments and social support.

With her plan in place, Emma took the courageous step to end the relationship. She moved to a new apartment and began rebuilding her life. Emma's story spread throughout Numeria, inspiring others to use math to evaluate and improve their own relationships.

Emma's story proved that math could be a powerful tool in escaping a bad relationship. Her innovative approach inspired others to embrace mathematical principles in their pursuit of healthier, happier lives. Numeria became a beacon of empowered individuals, demonstrating that a balanced life, guided by math, could lead to lasting happiness.

Compacting the Story

250 Words

In Numeria, mathematician Emma found herself trapped in a toxic relationship. She decided to use her mathematical skills to find a way out. Emma analyzed her relationship, assigning numerical values to aspects such as happiness, stress, and support.

She created the Relationship Health Equation:

$$R = \frac{(H - S + Su)}{T}$$

, where

R represented the relationship score,

H represented happiness,

S

represented stress,

S_u represented support, and

T represented time.

Emma tested her equation by applying it to her relationship. She tracked her happiness, stress, and support levels over several months and found the equation accurately predicted the decline in her relationship's health. The relationship score was consistently low, confirming her need to leave.

Encouraged by her findings, Emma devised a plan to escape. She used her mathematical skills to create a budget, ensuring she had enough resources to support herself. She calculated the optimal time to leave, considering factors such as work commitments and social support.

With her plan in place, Emma ended the relationship and moved to a new apartment. Her story spread throughout Numeria, inspiring others to use math to evaluate and improve their own relationships.

Emma's story proved math could be a powerful tool in escaping a bad relationship. Her approach inspired others to embrace mathematical principles in their pursuit of healthier, happier lives.

200 Words

In Numeria, mathematician Emma found herself trapped in a toxic relationship. She decided to use her mathematical skills to find a way out. Emma analyzed her relationship, assigning numerical values to aspects such as happiness, stress, and support.

She created the Relationship Health Equation:

$$R = (H - S + Su)T$$

, where

R represented the relationship score,

H represented happiness,

S represented stress,

Su represented support, and

T represented time.

Emma tested her equation by applying it to her relationship. She tracked her happiness, stress, and support levels over several months and found the equation accurately predicted the decline in her relationship's health. The relationship score was consistently low, confirming her need to leave.

Encouraged by her findings, Emma devised a plan to escape. She used her mathematical skills to create a budget, ensuring she had enough resources to support herself. She calculated the optimal time to leave, considering factors such as work commitments and social support.

With her plan in place, Emma ended the relationship and moved to a new apartment. Her story spread throughout Numeria, inspiring others to use math to evaluate and improve their own relationships.

Emma's story proved math could be a powerful tool in escaping a bad relationship.

150 Words

In Numeria, mathematician Emma found herself trapped in a toxic relationship. She decided to use her mathematical skills to find a way out. Emma analyzed her relationship, assigning numerical values to aspects such as happiness, stress, and support.

She created the Relationship Health Equation:

$R = (H - S$

$+ Su)T$, where

R represented the relationship score,

H represented happiness,

S represented stress,

Su represented support, and

T represented time.

Emma tested her equation by applying it to her relationship. She tracked her happiness, stress, and support levels over several months and found the equation accurately predicted the decline in her relationship's health. The relationship score was consistently low, confirming her need to leave.

Encouraged by her findings, Emma devised a plan to escape. She created a budget and calculated the optimal time to leave. With her plan in place, Emma ended the relationship and moved to a new apartment.

100 Words

In Numeria, mathematician Emma found herself trapped in a toxic relationship. She decided to use her mathematical skills to find a way out. Emma analyzed her relationship, assigning numerical values to aspects such as happiness, stress, and support.

She created the Relationship Health Equation:

$R = (H - S + S$

u

)

T

.

Emma tested her equation by applying it to her relationship. She tracked her happiness, stress, and support levels and found the equation accurately predicted the decline in her relationship's health. The relationship score was consistently low, confirming her need to leave.

Encouraged by her findings, Emma devised a plan to escape.

50 Words

In Numeria, mathematician Emma used math to escape a toxic relationship. She created the Relationship Health Equation:

R

=

(

H

−

S

+ Su) T.

Emma tested it and found the relationship score was low. Encouraged by her findings, Emma devised a plan to escape.

25 Words

Mathematician Emma created the Relationship Health Equation:

$R = (H - S + Su)T$

)
T
.

She used it to escape a toxic relationship.

20 Words

Mathematician Emma created the Relationship Health Equation. She used it to escape a toxic relationship.

15 Words

Emma created the Relationship Health Equation. She escaped a toxic relationship.

10 Words

Emma's equation helped her escape a toxic relationship.

7 Words

Emma's equation helped escape a toxic relationship.

5 Words

Equation helped escape toxic relationship.

4 Words

Equation escaped relationship.

3 Words

Escaped relationship.

2 Words

Relationship escaped.

1 Word

Escaped.

0 Words

(Silence)

Explanation

The story illustrates how math can be used to escape a bad relationship. By compacting the story, we distill its essence, showing how the core message remains powerful even in its simplest form.

Detailed Comments

Clarity: The story and its compacted versions maintain a clear message about the power of math in escaping a bad relationship.

Dynamism: The use of mathematical principles adds a dynamic element to the narrative.

Efficiency: The compacted versions efficiently convey the core message without losing its essence.

Effectiveness: Each version effectively communicates the relationship.

CHAPTER 26

Highlights the relationship of math how to win a major cooking contest:

Microfiction Story (299 words)

In the bustling city of Gourmandia, a mathematician named Leo was determined to win the prestigious Grand Culinary Contest. He believed that math could help him create the perfect dish. Leo began by analyzing the winning recipes from previous years, identifying patterns and common elements.

He created a mathematical model to quantify the factors that contributed to a winning dish, such as flavor balance, texture, and presentation. Leo called it the Culinary Success Equation: $C=(F+T+P)E$, where C represented the culinary score, F represented flavor, T represented texture, P represented presentation, and E represented effort.

Leo tested his equation by applying it to his own recipes. He meticulously measured ingredients, adjusted cooking times, and perfected plating techniques. He tracked his progress over several weeks and found that the equation accurately predicted the success of his dishes.

Encouraged by his success, Leo entered the Grand Culinary Contest with a dish that scored the highest on his equation. The judges were impressed by the perfect balance of flavors, the delightful texture, and the stunning presentation. Leo's dish stood out among the competition, and he was awarded the grand prize.

Leo's story proved that math could be a powerful tool in the culinary arts. His innovative approach inspired other chefs to embrace mathematical principles in their cooking. Gourmandia became a hub of culinary excellence, demonstrating that a balanced life, guided by math, could lead to delicious success.

Compacting the Story

250 Words

In Gourmandia, mathematician Leo sought to win the Grand Culinary Contest. He believed math could help him create the perfect dish. Leo analyzed winning recipes from previous years, identifying patterns and common elements.

He created the Culinary Success Equation: $C=(F+T+P)E$, where C represented the culinary score, F represented flavor, T represented texture, P represented presentation, and E represented effort.

Leo tested his equation by applying it to his own recipes. He meticulously measured ingredients, adjusted cooking times, and perfected plating techniques. He tracked his progress over several weeks and found the equation accurately predicted the success of his dishes.

Encouraged by his success, Leo entered the Grand Culinary Contest with a dish that scored the highest on his equation. The judges were impressed by the perfect balance of flavors, the delightful texture, and the stunning presentation. Leo's dish stood out among the competition, and he was awarded the grand prize.

Leo's story proved math could be a powerful tool in the culinary arts. His approach inspired other chefs to embrace mathematical principles in their cooking. Gourmandia became a hub of culinary excellence, demonstrating that a balanced life, guided by math, could lead to delicious success.

200 Words

In Gourmandia, mathematician Leo sought to win the Grand Culinary Contest. He believed math could help him create the perfect dish. Leo analyzed winning recipes from previous years, identifying patterns and common elements.

He created the Culinary Success Equation: $C=(F+T+P)E$, where C represented the culinary score, F represented flavor, T represented texture, P represented presentation, and E represented effort.

Leo tested his equation by applying it to his own recipes. He meticulously measured ingredients, adjusted cooking times, and perfected plating techniques. He tracked his progress over several weeks and found the equation accurately predicted the success of his dishes.

Encouraged by his success, Leo entered the Grand Culinary Contest with a dish that scored the highest on his equation. The judges were impressed by the perfect balance of flavors, the delightful texture, and the stunning presentation. Leo's dish stood out among the competition, and he was awarded the grand prize.

Leo's story proved math could be a powerful tool in the culinary arts.

150 Words

In Gourmandia, mathematician Leo sought to win the Grand Culinary Contest. He believed math could help him create the perfect dish. Leo analyzed winning recipes from previous years, identifying patterns and common elements.

He created the Culinary Success Equation: $C=(F+T+P)E$, where C represented the culinary score, F represented flavor, T represented texture, P represented presentation, and E represented effort.

Leo tested his equation by applying it to his own recipes. He meticulously measured ingredients, adjusted cooking times, and perfected plating techniques. He tracked his progress and found the equation accurately predicted the success of his dishes.

Encouraged by his success, Leo entered the Grand Culinary Contest with a dish that scored the highest on his equation. The judges were impressed, and Leo was awarded the grand prize.

Leo's story proved math could be a powerful tool in the culinary arts.

100 Words

In Gourmandia, mathematician Leo sought to win the Grand Culinary Contest. He believed math could help him create the perfect dish. Leo analyzed winning recipes from previous years, identifying patterns and common elements.

He created the Culinary Success Equation: $C=(F+T+P)E$.

Leo tested his equation by applying it to his own recipes. He meticulously measured ingredients, adjusted cooking times, and perfected plating techniques. He tracked his progress and found the equation accurately predicted the success of his dishes.

Encouraged by his success, Leo entered the contest and won the grand prize.

50 Words

In Gourmandia, mathematician Leo used math to win the Grand Culinary Contest. He created the Culinary Success Equation: $C=(F+T+P)E$. Leo tested it and won the grand prize.

25 Words

Mathematician Leo created the Culinary Success Equation: $C=(F+T+P)E$. He won the Grand Culinary Contest.

20 Words

Mathematician Leo created the Culinary Success Equation. He won the Grand Culinary Contest.

15 Words

Leo created the Culinary Success Equation. He won.

10 Words

Leo's equation won the culinary contest.

7 Words

Leo's equation won the contest.

5 Words

Equation won the contest.

4 Words

Equation won contest.

3 Words

Equation won.

2 Words

Equation won.

1 Word

Won.

0 Words

(Silence)

Explanation

The story illustrates how math can be used to win a major cooking contest. By compacting the story, we distill its essence, showing how the core message remains powerful even in its simplest form.

Detailed Comments

- **Clarity**: The story and its compacted versions maintain a clear message about the power of math in winning a cooking contest.

- **Dynamism**: The use of mathematical principles adds a dynamic element to the narrative.

- **Efficiency**: The compacted versions efficiently convey the core message without losing its essence.

- **Effectiveness**: Each version effectively communicates the relationship between math and culinary success.

- **Realism**: The story is grounded in realistic applications of math in cooking.

Series of Opinions

1. Math can be a powerful tool for culinary success.
2. Mathematical principles can simplify complex tasks.
3. Compacting stories can reveal their core essence.
4. Math can guide practical ways to achieve culinary excellence.
5. Practical math can inspire creativity in cooking.

Sonnets

Sonnet 1: The Culinary Success Equation

In Gourmandia, where flavors blend and play, A mathematician sought to find a way, To win the contest with a perfect dish, With math and logic, he would grant his wish.

He crafted formulas to light the way, To balance flavor, texture, love each day. The Culinary Success Equation was his key, To finding taste and living happily.

His town embraced his method, clear and bright, And found their plates were filled with pure delight. With math and wisdom, they did find their way, To culinary joy that lasts both night and day.

His story proves that numbers hold the key, To living life with taste and harmony.

Sonnet 2: Leo's Mathematical Quest

In Gourmandia, Leo's quest began, To find a way to help his fellow man. With math and data, he did seek to find, A method to bring peace to troubled mind.

He gathered experts, analysts, and more, To build a model, solid to the core. The Culinary Success Equation was their aim, A formula to bring flavor's gentle flame.

With careful thought and rigorous design, They crafted tools to help the taste align. The town embraced his method, clear and bright, And found their plates were filled with pure delight.

His tale reminds us all of math's great power, To turn the tide and bring a brighter hour.

Epic Poem: The Quest for Flavor

In Gourmandia, a city bright and grand, A mathematician took a daring stand. Leo was his name, a mind so keen, He sought to find a way to intervene.

The search for flavor that plagued his heart and mind, He sought to ease with formulas to find. He crafted tools to light the way, To balance flavor, texture, love each day.

The Culinary Success Equation was his key, To finding taste and living happily. His town embraced his method, clear and bright, And found their plates were filled with pure delight.

Through trials and tests, they forged a path so bright, Their formula a beacon in the night. Leo's story spread, a legend far and wide, Of how math's power turned the culinary tide.

***

Need a homestead lifestyle:
Microfiction Story (299 words)

In the quiet countryside of Arborville, a young woman named Clara decided to embrace a homestead lifestyle. She knew that mastering vital skills like gardening, carpentry, and animal husbandry would be essential. Clara, a former math teacher, saw an opportunity to apply her mathematical knowledge to her new life.

Clara began with gardening. She used geometry to design efficient garden beds, ensuring optimal sunlight and water distribution. She calculated the ideal spacing for each plant, maximizing yield while minimizing resource use. Her garden flourished, providing fresh produce for her family and neighbors.

Next, Clara tackled carpentry. She used algebra to determine the precise measurements for building a chicken coop. By applying the Pythagorean theorem, she ensured the structure was sturdy and well-balanced. The coop became a safe haven for her chickens, who provided eggs and meat.

Animal husbandry was her final challenge. Clara used statistics to track the health and productivity of her livestock. She calculated feed ratios and monitored growth rates, ensuring her animals thrived. Her farm became a model of efficiency and sustainability.

Clara's mathematical approach to homesteading caught the attention of her community. She began teaching workshops, sharing her knowledge and inspiring others to embrace math in their daily lives. Arborville thrived, becoming a beacon of sustainable living.

Through her journey, Clara demonstrated that math was not just a subject confined to classrooms but a practical tool that could transform lives. Her story became a testament to the power of numbers in creating a harmonious and productive homestead.

Compacting the Story

250 Words

In Arborville, Clara embraced a homestead lifestyle, applying her math skills to gardening, carpentry, and animal husbandry. She used geometry to design efficient garden beds, ensuring optimal sunlight and water distribution. Her garden flourished, providing fresh produce for her family and neighbors.

For carpentry, Clara used algebra to determine precise measurements for building a chicken coop. By applying the

Pythagorean theorem, she ensured the structure was sturdy and well-balanced. The coop became a safe haven for her chickens, who provided eggs and meat.

In animal husbandry, Clara used statistics to track the health and productivity of her livestock. She calculated feed ratios and monitored growth rates, ensuring her animals thrived. Her farm became a model of efficiency and sustainability.

Clara's mathematical approach to homesteading caught the attention of her community. She began teaching workshops, sharing her knowledge and inspiring others to embrace math in their daily lives. Arborville thrived, becoming a beacon of sustainable living.

Through her journey, Clara demonstrated that math was not just a subject confined to classrooms but a practical tool that could transform lives. Her story became a testament to the power of numbers in creating a harmonious and productive homestead.

200 Words

In Arborville, Clara embraced a homestead lifestyle, applying her math skills to gardening, carpentry, and animal husbandry. She used geometry to design efficient garden beds, ensuring optimal sunlight and water distribution. Her garden flourished, providing fresh produce for her family and neighbors.

For carpentry, Clara used algebra to determine precise measurements for building a chicken coop. By applying the Pythagorean theorem, she ensured the structure was sturdy and well-balanced. The coop became a safe haven for her chickens, who provided eggs and meat.

In animal husbandry, Clara used statistics to track the health and productivity of her livestock. She calculated feed ratios and monitored growth rates, ensuring her animals thrived. Her farm became a model of efficiency and sustainability.

Clara's mathematical approach to homesteading caught the attention of her community. She began teaching workshops, sharing her knowledge and inspiring others to embrace math in their daily lives. Arborville thrived, becoming a beacon of sustainable living.

Through her journey, Clara demonstrated that math was not just a subject confined to classrooms but a practical tool that could transform lives. Her story became a testament to the power of numbers in creating a harmonious and productive homestead.

150 Words

In Arborville, Clara embraced a homestead lifestyle, applying her math skills to gardening, carpentry, and animal husbandry. She used geometry to design efficient garden beds, ensuring optimal sunlight and water distribution. Her garden flourished, providing fresh produce for her family and neighbors.

For carpentry, Clara used algebra to determine precise measurements for building a chicken coop. By applying the Pythagorean theorem, she ensured the structure was sturdy and well-balanced. The coop became a safe haven for her chickens, who provided eggs and meat.

In animal husbandry, Clara used statistics to track the health and productivity of her livestock. She calculated feed ratios and monitored growth rates, ensuring her animals thrived. Her farm became a model of efficiency and sustainability.

Clara's mathematical approach to homesteading caught the attention of her community. She began teaching workshops, sharing her knowledge and inspiring others. Arborville thrived, becoming a beacon of sustainable living.

100 Words

In Arborville, Clara embraced a homestead lifestyle, applying her math skills to gardening, carpentry, and animal husbandry. She used geometry to design efficient garden beds, ensuring optimal sunlight and water distribution. Her garden flourished.

For carpentry, Clara used algebra to determine precise measurements for building a chicken coop. By applying the Pythagorean theorem, she ensured the structure was sturdy and well-balanced. The coop became a safe haven for her chickens.

In animal husbandry, Clara used statistics to track the health and productivity of her livestock. Her farm became a model of efficiency and sustainability. Clara's story demonstrated math's practical power.

50 Words

In Arborville, Clara used math for gardening, carpentry, and animal husbandry. Geometry optimized her garden, algebra built a sturdy chicken coop, and statistics tracked livestock health. Her farm thrived, inspiring the community. Clara's story showed math's practical power.

25 Words

Clara used math for gardening, carpentry, and animal husbandry. Her farm thrived, inspiring the community. Clara's story showed math's practical power.

20 Words

Clara used math for gardening, carpentry, and animal husbandry. Her farm thrived, inspiring the community.

15 Words

Clara used math for homesteading. Her farm thrived, inspiring the community.

10 Words

Clara used math. Her farm thrived, inspiring others.

7 Words

Clara's math thrived, inspiring others.

5 Words

Clara's math inspired others.

4 Words

Math inspired others.

3 Words

Math inspired.

2 Words

Math inspired.

1 Word

Inspired.

0 Words

(Silence)

Explanation

The story illustrates how math can be used as a practical tool for learning vital homestead skills. By compacting the story, we distill its essence, showing how the core message remains powerful even in its simplest form.

Detailed Comments

- **Clarity**: The story and its compacted versions maintain a clear message about the practical power of math.

- **Dynamism**: The use of mathematical applications adds a dynamic element to the narrative.

- **Efficiency**: The compacted versions efficiently convey the core message without losing its essence.

- **Effectiveness**: Each version effectively communicates the relationship between math and homesteading.

- **Realism**: The story is grounded in realistic applications of math in everyday life.

Series of Opinions

1. Math can be a powerful tool for learning vital skills.
2. Mathematical applications can simplify complex tasks.
3. Compacting stories can reveal their core essence.
4. Math can guide practical ways to achieve efficiency and sustainability.
5. Practical math can inspire communities.

Poetic Math Functions

1. **Clarity**: $f(x)=clear(x)$
2. **Dynamism**: $g(x)=dynamic(x)$
3. **Efficiency**: $h(x)=efficient(x)$
4. **Effectiveness**: $i(x)=effective(x)$
5. **Realism**: $j(x)=realistic(x)$

Combining these functions: $$S(x) = f(x) + g(x) + h(x) + i(x) + j(x)$$

This function S(x) represents an exceptional story that embodies clarity, dynamism, efficiency, effectiveness, and realism.

***

Highlights the relationship of math and how to make a man happy:

Microfiction Story (299 words)

In the bustling city of Numeria, a mathematician named Clara was determined to make her partner, Leo, truly happy. She believed that math could help her understand his needs and desires. Clara began by creating a list of factors that contributed to Leo's happiness, such as quality time, appreciation, and shared interests.

She assigned numerical values to each factor and developed a mathematical model to evaluate their impact on Leo's happiness. Clara called it the Happiness Optimization Equation: $H=(Q+A+S)T$, where H represented happiness, Q represented quality time, A represented appreciation, S represented shared interests, and T represented time spent together.

Clara tested her equation by applying it to their daily lives. She made sure to spend quality time with Leo, express her appreciation, and engage in activities they both enjoyed. She tracked Leo's happiness levels over several weeks and found that the equation accurately predicted his overall well-being.

Encouraged by her success, Clara shared her findings with her friends. They began to use the Happiness Optimization Equation to improve their own relationships, making adjustments to meet their partners' needs. The city of Numeria thrived, with residents experiencing increased happiness and stronger connections.

Clara's story proved that math could be a powerful tool in understanding and enhancing relationships. Her innovative approach inspired others to embrace mathematical principles in their pursuit of happiness. Numeria became a beacon of joyful partnerships, demonstrating that a balanced life, guided by math, could lead to lasting happiness.

Compacting the Story

250 Words

In Numeria, mathematician Clara sought to make her partner, Leo, truly happy. She believed math could help her understand his needs and desires. Clara created a list of factors that contributed to Leo's happiness, such as quality time, appreciation, and shared interests.

She assigned numerical values to each factor and developed the Happiness Optimization Equation: $H=(Q+A+S)T$, where H represented happiness, Q represented quality time, A represented appreciation, S represented shared interests, and T represented time spent together.

Clara tested her equation by applying it to their daily lives. She made sure to spend quality time with Leo, express her appreciation, and engage in activities they both enjoyed. She tracked Leo's happiness levels over several weeks and found the equation accurately predicted his well-being.

Encouraged by her success, Clara shared her findings with friends. They used the Happiness Optimization Equation to improve their own relationships, meeting their partners' needs. Numeria thrived, with residents experiencing increased happiness and stronger connections.

Clara's story proved math could be a powerful tool in understanding and enhancing relationships. Her approach inspired others to embrace mathematical principles in their pursuit of happiness. Numeria became a beacon of joyful partnerships, demonstrating that a balanced life, guided by math, could lead to lasting happiness.

200 Words

In Numeria, mathematician Clara sought to make her partner, Leo, truly happy. She believed math could help her understand his needs and desires. Clara created a list of factors that contributed to Leo's happiness, such as quality time, appreciation, and shared interests.

She assigned numerical values to each factor and developed the Happiness Optimization Equation: $H=(Q+A+S)T$, where H represented happiness, Q represented quality time, A represented appreciation, S represented shared interests, and T represented time spent together.

Clara tested her equation by applying it to their daily lives. She made sure to spend quality time with Leo, express her appreciation, and engage in activities they both enjoyed. She tracked Leo's happiness levels over several weeks and found the equation accurately predicted his well-being.

Encouraged by her success, Clara shared her findings with friends. They used the Happiness Optimization Equation to improve their own relationships, meeting their partners' needs. Numeria thrived, with residents experiencing increased happiness and stronger connections.

Clara's story proved math could be a powerful tool in understanding and enhancing relationships.

150 Words

In Numeria, mathematician Clara sought to make her partner, Leo, truly happy. She believed math could help her understand his needs and desires. Clara created a list of factors that contributed to Leo's happiness, such as quality time, appreciation, and shared interests.

She assigned numerical values to each factor and developed the Happiness Optimization …

***

Highlights the relationship of math and how get rid of bad thoughts:

Microfiction Story (299 words)

In the bustling city of Numeria, a mathematician named Clara was plagued by bad thoughts. She believed that math could help her find a way to clear her mind. Clara began by analyzing her thoughts and identifying patterns in her negative thinking.

She created a mathematical model to quantify the impact of different activities on her mental state. Clara called it the Thought Clarity Equation: $T=(P+M+R)S$, where T represented thought clarity, P represented positive activities, M represented mindfulness practices, R represented relaxation, and S represented stress.

Clara tested her equation by applying it to her daily routine. She balanced her time between positive activities, mindfulness practices, and relaxation, ensuring that she minimized stress. She tracked her mental state over several weeks and found that the equation accurately predicted her overall well-being.

Encouraged by her success, Clara shared her findings with her friends and colleagues. People began to use the Thought Clarity Equation to guide their daily lives, making adjustments to achieve a balanced and clear mind. The city of Numeria thrived, with residents experiencing increased happiness and mental clarity.

Clara's story proved that math could be a powerful tool in clearing bad thoughts. Her innovative approach inspired others to embrace mathematical principles in their pursuit of mental well-being. Numeria became a beacon of mental clarity, demonstrating that a balanced life, guided by math, could lead to lasting peace of mind.

Compacting the Story

250 Words

In Numeria, mathematician Clara was plagued by bad thoughts. She believed math could help her clear her mind. Clara analyzed her thoughts and identified patterns in her negative thinking.

She created the Thought Clarity Equation: $T=(P+M+R)S$, where T represented thought clarity, P represented positive activities, M represented mindfulness practices, R represented relaxation, and S represented stress.

Clara tested her equation by applying it to her daily routine. She balanced her time between positive activities, mindfulness practices, and relaxation, minimizing stress. She tracked her mental state over several weeks and found the equation accurately predicted her well-being.

Encouraged by her success, Clara shared her findings with friends and colleagues. People used the Thought Clarity Equation to guide their lives, achieving a balanced and clear mind. Numeria thrived, with residents experiencing increased happiness and mental clarity.

Clara's story proved math could be a powerful tool in clearing bad thoughts. Her approach inspired others to embrace mathematical principles in their pursuit of mental well-being. Numeria became a beacon of mental clarity, demonstrating that a balanced life, guided by math, could lead to lasting peace of mind.

200 Words

In Numeria, mathematician Clara was plagued by bad thoughts. She believed math could help her clear her mind. Clara analyzed her thoughts and identified patterns in her negative thinking.

She created the Thought Clarity Equation: $T=(P+M+R)S$, where T represented thought clarity, P represented positive activities, M represented mindfulness practices, R represented relaxation, and S represented stress.

Clara tested her equation by applying it to her daily routine. She balanced her time between positive activities, mindfulness practices, and relaxation, minimizing stress. She tracked her mental state over several weeks and found the equation accurately predicted her well-being.

Encouraged by her success, Clara shared her findings with friends and colleagues. People used the Thought Clarity Equation to guide their lives, achieving a balanced and clear mind. Numeria thrived, with residents experiencing increased happiness and mental clarity.

Clara's story proved math could be a powerful tool in clearing bad thoughts. Her approach inspired others to embrace mathematical principles in their pursuit of mental well-being.

150 Words

In Numeria, mathematician Clara was plagued by bad thoughts. She believed math could help her clear her mind. Clara analyzed her thoughts and identified patterns in her negative thinking.

She created the Thought Clarity Equation: $T=(P+M+R)S$, where T represented thought clarity, P represented positive activities, M represented mindfulness practices, R represented …

***

Highlights the relationship of math and create an excellent blockbuster high concept adventure

Microfiction Story (299 words)

In the bustling city of Numeria, a mathematician named Leo was determined to create an excellent blockbuster high-concept adventure. He believed that math could help him craft the perfect story. Leo began by analyzing the structure of successful adventure films, identifying patterns and common elements.

He created a mathematical model to quantify the factors that contributed to a blockbuster hit, such as plot twists, character development, and pacing. Leo called it the Adventure Success Equation: $A=(P+C+T)E$, where A represented the adventure score, P represented plot twists, C represented character development, T represented pacing, and E represented effort.

Leo tested his equation by applying it to his own screenplay. He meticulously crafted each scene, ensuring that the plot twists were surprising, the characters were well-developed, and the pacing was perfect. He tracked his progress over several months and found that the equation accurately predicted the success of his story.

Encouraged by his success, Leo pitched his screenplay to a major film studio. The producers were impressed by the perfect balance of excitement, depth, and engagement. Leo's film stood out among the competition, and it was greenlit for production.

When the film was released, it became an instant blockbuster, captivating audiences worldwide. Critics praised its perfect pacing, engaging characters, and thrilling plot twists. Leo's story proved that math could be a powerful tool in the world of filmmaking. His innovative approach inspired other filmmakers to embrace mathematical principles in their storytelling. Numeria became a hub of cinematic excellence, demonstrating that a balanced life, guided by math, could lead to blockbuster success.

Compacting the Story

200 Words

In Numeria, mathematician Leo sought to create a blockbuster high-concept adventure. He believed math could help him craft the perfect story. Leo analyzed successful adventure films, identifying patterns and common elements.

He created the Adventure Success Equation: $A=(P+C+T)E$, where A represented the adventure score, P represented plot twists, C represented character development, T represented pacing, and E represented effort.

Leo tested his equation by applying it to his screenplay. He meticulously crafted each scene, ensuring plot twists were surprising, characters well-developed, and pacing perfect. He tracked his progress and found the equation accurately predicted the success of his story.

Encouraged by his success, Leo pitched his screenplay to a major film studio. The producers were impressed by the balance of excitement, depth, and engagement. Leo's film was greenlit for production.

When released, the film became an instant blockbuster, captivating audiences worldwide. Critics praised its pacing, characters, and plot twists. Leo's story proved math could be a powerful tool in filmmaking. His approach inspired other filmmakers to embrace mathematical principles in storytelling.

150 Words

In Numeria, mathematician Leo sought to create a blockbuster high-concept adventure. He believed math could help him craft the perfect story. Leo analyzed successful adventure films, identifying patterns and common elements.

He created the Adventure Success Equation: $A=(P+C+T)E$, where A represented the adventure score, P represented plot twists, C represented character development, T represented pacing, and E represented effort.

Leo tested his equation by applying it to his screenplay. He meticulously crafted each scene, ensuring plot twists were surprising, characters well-developed, and pacing perfect.

He tracked his progress and found the equation accurately predicted the success of his story.

Encouraged by his success, Leo pitched his screenplay to a major film studio. The producers were impressed, and the film was greenlit for production.

When released, the film became an instant blockbuster, captivating audiences worldwide.

100 Words

In Numeria, mathematician Leo sought to create a blockbuster high-concept adventure. He believed math could help him craft the perfect story. Leo analyzed successful adventure films, identifying patterns and common elements.

He created the Adventure Success Equation: $A=(P+C+T)E$.

Leo tested his equation by applying it to his screenplay. He meticulously crafted each scene, ensuring plot twists were surprising, characters well-developed, and pacing perfect. He tracked his progress and found the equation accurately predicted the success of his story.

Encouraged by his success, Leo pitched his screenplay, and it was greenlit for production.

50 Words

In Numeria, mathematician Leo used math to create a blockbuster adventure. He created the Adventure Success Equation: A=(P+C+T)E. Leo tested it and pitched his screenplay. It was greenlit for production and became an instant blockbuster.

25 Words

Mathematician Leo created the Adventure Success Equation: $A =$

CHAPTER 27

Highlights the relationship of math and create the perfect pillow and bed:

Microfiction Story (299 words)

In the bustling city of Numeria, a mathematician named Clara was determined to create the perfect pillow and bed. She believed that math could help her achieve the ultimate comfort. Clara began by analyzing the factors that contributed to a good night's sleep, such as support, pressure relief, and temperature regulation.

She assigned numerical values to each factor and developed a mathematical model to optimize the design of the pillow and bed. Clara called it the Sleep Optimization Equation: $S=(Su+Pr+T)C$, where S represented sleep quality, Su represented support, Pr represented pressure relief, T represented temperature regulation, and C represented comfort.

Clara tested her equation by applying it to various pillow and bed designs. She collected data on each design and calculated the sleep quality scores. She tried out the top-scoring designs to see if they matched her expectations.

Clara found that the equation accurately predicted the comfort and support of each design.

Encouraged by her success, Clara shared her findings with her friends and colleagues. People began to use the Sleep Optimization Equation to guide their decisions on pillows and beds, making adjustments to find the perfect combination for their needs. The city of Numeria thrived, with residents experiencing improved sleep and overall well-being.

Clara's story proved that math could be a powerful tool in creating the perfect pillow and bed. Her innovative approach inspired others to embrace mathematical principles in their pursuit of comfort. Numeria became a beacon of restful nights, demonstrating that a balanced life, guided by math, could lead to lasting happiness.

Compacting the Story

200 Words

In Numeria, mathematician Clara sought to create the perfect pillow and bed. She believed math could help her achieve ultimate comfort. Clara analyzed factors contributing to a good night's sleep, such as support, pressure relief, and temperature regulation.

She assigned numerical values to each factor and developed the Sleep Optimization Equation: $S=(Su+Pr+T)C$, where S represented sleep quality, Su represented support, Pr represented pressure relief, T represented temperature regulation, and C represented comfort.

Clara tested her equation by applying it to various pillow and bed designs. She collected data on each design and calculated the sleep quality scores. She tried out the top-scoring designs to see if they matched her expectations. Clara found that the equation accurately predicted the comfort and support of each design.

Encouraged by her success, Clara shared her findings with friends and colleagues. People used the Sleep Optimization Equation to guide their decisions on pillows and beds, finding the perfect combination for their needs. Numeria thrived, with residents experiencing improved sleep and overall well-being.

Clara's story proved math could be a powerful tool in creating the perfect pillow and bed.

150 Words

In Numeria, mathematician Clara sought to create the perfect pillow and bed. She believed math could help her achieve ultimate comfort. Clara analyzed factors contributing to a good night's sleep, such as support, pressure relief, and temperature regulation.

She assigned numerical values to each factor and developed the Sleep Optimization Equation: $S=(Su+Pr+T)C$, where S represented sleep quality, Su represented support, Pr represented pressure relief, T represented temperature regulation, and C represented comfort.

Clara tested her equation by applying it to various pillow and bed designs. She collected data on each design and calculated the sleep quality scores. Clara found that the equation accurately predicted the comfort and support of each design.

Encouraged by her success, Clara shared her findings. Numeria thrived, with residents experiencing improved sleep and overall well-being.

100 Words

In Numeria, mathematician Clara sought to create the perfect pillow and bed. She believed math could help her achieve ultimate comfort. Clara analyzed factors contributing to a good night's sleep, such as support, pressure relief, and temperature regulation.

She assigned numerical values to each factor and developed the Sleep Optimization Equation: $S=(Su+Pr+T)C$.

Clara tested her equation by applying it to various pillow and bed designs. She collected data on each design and calculated the sleep quality scores. Clara found that the equation accurately predicted the comfort and support of each design.

Encouraged by her success, Clara shared her findings.

50 Words

In Numeria, Clara used math to create the perfect pillow and bed. She created the Sleep Optimization Equation: $S=(Su+Pr+T)C$. Clara tested it and shared her findings. Numeria thrived, with residents experiencing improved sleep.

25 Words

Clara created the Sleep Optimization Equation: $S=(Su+Pr+T)C$. Numeria thrived, with residents experiencing improved sleep.

20 Words

Clara created the Sleep Optimization Equation. Numeria thrived, with residents experiencing improved sleep.

15 Words

Clara created the Sleep Optimization Equation. Numeria thrived.

10 Words

Clara's equation improved sleep. Numeria thrived.

7 Words

Clara's equation improved sleep quality.

5 Words

Equation improved sleep quality.

4 Words

Equation improved sleep.

3 Words

Improved sleep.

2 Words

Sleep improved.

1 Word

Sleep.

0 Words

(Silence)

Explanation

The story illustrates how math can be used to create the perfect pillow and bed. By compacting the story, we distill its essence, showing how the core message remains powerful even in its simplest form.

Detailed Comments

- **Clarity**: The story and its compacted versions maintain a clear message about the power of math in improving sleep quality.
- **Dynamism**: The use of mathematical principles adds a dynamic element to the narrative.
- **Efficiency**: The compacted versions efficiently convey the core message without losing its essence.
- **Effectiveness**: Each version effectively communicates the relationship between math and sleep optimization.
- **Realism**: The story is grounded in realistic applications of math in personal comfort.

Series of Opinions

1. Math can be a powerful tool for improving sleep quality.
2. Mathematical principles can simplify complex tasks.

3. Compacting stories can reveal their core essence.
4. Math can guide practical ways to achieve personal comfort.
5. Practical math can inspire well-being in personal life.

Sonnet: The Sleep Optimization Equation

***

Highlights the relationship of math and create an excellent high concept adventure to solve the Middle East crisis:

Microfiction Story (299 words)

In the heart of the Middle East, a brilliant mathematician named Layla was determined to find a solution to the region's long-standing conflicts. She believed that math could provide a path to peace. Layla gathered a team of experts in various fields, including political science, sociology, and economics, to develop a mathematical model for conflict resolution.

They began by analyzing historical data on conflicts, identifying patterns and common factors. Layla created a mathematical model to quantify the factors that contributed to peace, such as economic stability, social cohesion, and political representation. She called it the Peace Equation:

$P=(E+S+R)C$, where P represented peace, E represented economic stability, S represented social cohesion, R represented political representation, and C represented conflict intensity.

Layla's team tested the equation by applying it to various scenarios in the Middle East. They found that the equation accurately predicted the outcomes of different peace initiatives. Encouraged by their success, they presented their findings to regional leaders, who were impressed by the mathematical rigor and practical application of the Peace Equation.

The leaders implemented policies based on the equation, focusing on economic development, social programs, and inclusive governance. Over time, the region experienced a significant reduction in conflict and an increase in stability and prosperity.

Layla's story proved that math could be a powerful tool in solving complex geopolitical issues. Her innovative approach inspired others to embrace mathematical principles in their pursuit of peace. The Middle East became a beacon of hope, demonstrating that a balanced life, guided by math, could lead to lasting harmony.

Compacting the Story

200 Words

In the Middle East, mathematician Layla sought to find a solution to the region's conflicts. She believed math could provide a path to peace. Layla gathered experts in political science, sociology, and economics to develop a mathematical model for conflict resolution.

They analyzed historical data on conflicts, identifying patterns and common factors. Layla created the Peace Equation: $P=(E+S+R)C$, where P represented peace, E represented economic stability, S represented social cohesion, R represented political representation, and C represented conflict intensity.

Layla's team tested the equation by applying it to various scenarios in the Middle East. They found that the equation accurately predicted the outcomes of different peace initiatives. Encouraged by their success, they presented their findings to regional leaders, who implemented policies based on the equation.

The region experienced a significant reduction in conflict and an increase in stability and prosperity. Layla's story proved that math could be a powerful tool in solving complex geopolitical issues. Her approach inspired others to embrace mathematical principles in their pursuit of peace.

150 Words

In the Middle East, mathematician Layla sought to find a solution to the region's conflicts. She believed math could provide a path to peace. Layla gathered experts in political science, sociology, and economics to develop a mathematical model for conflict resolution.

They analyzed historical data on conflicts, identifying patterns and common factors. Layla created the Peace Equation: $P=(E+S+R)C$, where P represented peace, E represented economic stability, S represented social cohesion, R represented political representation, and C represented conflict intensity.

Layla's team tested the equation by applying it to various scenarios in the Middle East. They found that the equation accurately predicted the outcomes of different peace initiatives. Regional leaders implemented policies based on the equation, leading to a reduction in conflict and an increase in stability and prosperity.

Layla's story proved that math could be a powerful tool in solving complex geopolitical issues.

100 Words

In the Middle East, mathematician Layla sought to find a solution to the region's conflicts. She believed math could provide a path to peace. Layla gathered experts to develop a mathematical model for conflict resolution.

They analyzed historical data on conflicts and created the Peace Equation: $P=(E+S+R)C$.

Layla's team tested the equation and found it accurately predicted the outcomes of peace initiatives. Regional leaders implemented policies based on the equation, leading to a reduction in conflict and an increase in stability and prosperity.

Layla's story proved that math could be a powerful tool in solving geopolitical issues.

50 Words

In the Middle East, mathematician Layla used math to find a solution to conflicts. She created the Peace Equation: $P=(E+S+R)C$. Regional leaders implemented policies based on the equation, leading to stability and prosperity.

25 Words

Mathematician Layla created the Peace Equation: P=(E+S+R)C. Regional leaders implemented policies, leading to stability and prosperity.

20 Words

Mathematician Layla created the Peace Equation. Regional leaders implemented policies, leading to stability and prosperity.

15 Words

Layla created the Peace Equation. Leaders implemented policies, leading to prosperity.

10 Words

Layla's equation led to stability and prosperity.

7 Words

Layla's equation led to prosperity.

5 Words

Equation led to prosperity.

4 Words

Equation led prosperity.

3 Words

Equation led.

2 Words

Equation led.

1 Word

Equation.

0 Words

(Silence)

Explanation

The story illustrates how math can be used to solve the Middle East crisis. By compacting the story, we distill its essence, showing how the core message remains powerful even in its simplest form.

Detailed Comments

- **Clarity**: The story and its compacted versions maintain a clear message about the power of math in solving geopolitical issues.
- **Dynamism**: The use of mathematical principles adds a dynamic element to the narrative.
- **Efficiency**: The compacted versions efficiently convey the core message without losing its essence.

- **Effectiveness**: Each version effectively communicates the relationship between math and conflict resolution.
- **Realism**: The story is grounded in realistic applications of math in geopolitical decisions.

Series of Opinions

1. Math can be a powerful tool for solving geopolitical issues.
2. Mathematical principles can simplify complex tasks.
3. Compacting stories can reveal their core essence.
4. Math can guide practical ways to achieve peace and stability.
5. Practical math can inspire hope in conflict resolution.

Sonnets

Sonnet 1: The Peace Equation

In the Middle East, where conflicts rage and roar, A mathematician sought to find a cure. With numbers, data, formulas, and more, She aimed to bring a peace that would endure.

She crafted equations to light the way, To balance economics, love each day. The Peace Equation was her key, To finding harmony and living free.

Her region embraced her method, clear and bright, And found their lives were filled with pure delight. With math and wisdom, they did find their way, To peace that lasts both night and day.

Her story proves that numbers hold the key, To living life with joy and harmony.

Sonnet 2: Layla's Mathematical Quest

In the …

**

Sonnets

Sonnet 1: The Literary Analysis Equation

In Literaria, where words and numbers blend, A mathematician sought a way to mend, Her love for stories, formulas, and more, To find the meaning she could not ignore.

She crafted equations to light the way, To balance plot, and themes, and love each day. The Literary Analysis Equation was her key, To finding truth and living happily.

Her town embraced her method, clear and bright, And found their tales were filled with pure delight. With math and wisdom, they did find their way, To literary joy that lasts both night and day.

Her story proves that numbers hold the key, To living life with words and harmony.

Sonnet 2: Clara's Mathematical Quest

In Literaria, Clara's quest began, To find a way to help her fellow man. With math and data, she did seek to find, A method to bring peace to troubled mind.

She gathered experts, analysts, and more, To build a model, solid to the core. The Literary Analysis Equation was their aim, A formula to bring story's gentle flame.

With careful thought and rigorous design, They crafted tools to help the words align. The town embraced her method, clear and bright, And found their tales were filled with pure delight.

Her tale reminds us all of math's great power, To turn the tide and bring a brighter hour.

Epic Poems

Epic Poem 1: The Quest for Story

In Literaria, a city bright and grand, A mathematician took a daring stand. Clara was her name, a mind so keen, She sought to find a way to intervene.

The search for meaning that plagued her heart and mind, She sought to ease with formulas to find. She crafted tools to light the way, To balance plot, and themes, and love each day.

The Literary Analysis Equation was her key, To finding truth and living happily. Her town embraced her method, clear and bright, And found their tales were filled with pure delight.

Through trials and tests, they forged a path so bright, Their formula a beacon in the night. Clara's story spread, a legend far and wide, Of how math's power turned the literary tide.

Epic Poem 2: Clara's Triumph

In Literaria, where stories come alive, A mathematician sought to help them thrive. Clara was her name, a mind so bright, She sought to bring the hidden truths to light.

With math and data, she did seek to find, A method to bring peace to troubled mind. She crafted tools to light the way, To balance plot, and themes, and love each day.

The Literary Analysis Equation was her key, To finding truth and living happily. Her town embraced her method, clear and bright, And found their tales were filled with pure delight.

Through trials and tests, they forged a path so bright, Their formula a beacon in the night. Clara's story spread, a legend far and wide, Of how math's power turned the literary tide.

Folk Song for Guitar: Clara's Equation

(Verse 1) In the town of Literaria, where stories come alive, A mathematician named Clara sought to help them thrive. With numbers and equations, she sought to find the way, To bring the hidden meanings to the light of day.

(Chorus) Oh, Clara's equation, lighting up the night, Bringing truth and wisdom, making stories bright. With math and love together, she …

***

Highlights the relationship of math and understand an unknown language:

Microfiction Story (299 words)

In the ancient city of Numeria, a mathematician named Leo discovered a mysterious manuscript written in an unknown language. Determined to decipher it, Leo believed that math could unlock its secrets. He began by analyzing the symbols and patterns in the text, identifying recurring sequences and structures.

Leo created a mathematical model to quantify the frequency and distribution of the symbols. He called it the Language Deciphering Equation: $L=(S+P+F)C$, where L represented the language score, S represented symbol frequency, P represented pattern recognition, F represented frequency analysis, and C represented context.

Leo tested his equation by applying it to the manuscript. He meticulously mapped out the symbols, comparing them to known languages and using his model to identify potential meanings. Over time, he began to recognize words and phrases, slowly piecing together the manuscript's message.

Encouraged by his progress, Leo shared his findings with linguists and historians. They used the Language Deciphering Equation to guide their own research, making adjustments to decode other ancient texts. The city of Numeria thrived, with scholars uncovering lost knowledge and preserving their heritage.

Leo's story proved that math could be a powerful tool in understanding unknown languages. His innovative approach inspired others to embrace mathematical principles in their linguistic pursuits. Numeria became a beacon of intellectual discovery, demonstrating that a balanced life, guided by math, could lead to lasting enlightenment.

Compacting the Story

200 Words

In Numeria, mathematician Leo discovered a mysterious manuscript in an unknown language. He believed math could unlock its secrets. Leo analyzed the symbols and patterns, identifying recurring sequences.

He created the Language Deciphering Equation: $L=(S+P+F)C$, where L represented the language score, S represented symbol frequency, P represented pattern recognition, F represented frequency analysis, and C represented context.

Leo tested his equation on the manuscript. He mapped out the symbols, comparing them to known languages and using his model to identify potential meanings. Over time, he recognized words and phrases, piecing together the manuscript's message.

Encouraged by his progress, Leo shared his findings with linguists and historians. They used the Language

Deciphering Equation to decode other ancient texts. Numeria thrived, with scholars uncovering lost knowledge and preserving their heritage.

Leo's story proved math could be a powerful tool in understanding unknown languages. His approach inspired others to embrace mathematical principles in their linguistic pursuits. Numeria became a beacon of intellectual discovery.

150 Words

In Numeria, mathematician Leo discovered a mysterious manuscript in an unknown language. He believed math could unlock its secrets. Leo analyzed the symbols and patterns, identifying recurring sequences.

He created the Language Deciphering Equation: $L=(S+P+F)C$, where L represented the language score, S represented symbol frequency, P represented pattern recognition, F represented frequency analysis, and C represented context.

Leo tested his equation on the manuscript. He mapped out the symbols, comparing them to known languages and using his model to identify potential meanings. Over time, he recognized words and phrases, piecing together the manuscript's message.

Encouraged by his progress, Leo shared his findings with linguists and historians. They used the Language Deciphering Equation to decode other ancient texts. Numeria thrived, with scholars uncovering lost knowledge and preserving their heritage.

Leo's story proved math could be a powerful tool in understanding unknown languages.

100 Words

In Numeria, mathematician Leo discovered a mysterious manuscript in an unknown language. He believed math could unlock its secrets. Leo analyzed the symbols and patterns, identifying recurring sequences.

He created the Language Deciphering Equation: $L=(S+P+F)C$.

Leo tested his equation on the manuscript. He mapped out the symbols, comparing them to known languages and using his model to identify potential meanings. Over time, he recognized words and phrases, piecing together the manuscript's message.

Encouraged by his progress, Leo shared his findings. Numeria thrived, with scholars uncovering lost knowledge.

50 Words

In Numeria, mathematician Leo used math to decipher a mysterious manuscript. He created …

CHAPTER 28

Highlights the relationship of math to music theory:

Microfiction Story (299 words)

In the harmonious city of Melodia, a mathematician named Leo was fascinated by the relationship between math and music theory. He believed that mathematical principles could unlock new dimensions in music composition. Leo began by analyzing the structure of classical compositions, identifying patterns in rhythm, harmony, and melody.

He created a mathematical model to quantify these elements, assigning values to different aspects of a musical piece. Leo called it the Music Harmony Equation: $M=(R+H+M)C$, where M represented the musical score, R represented rhythm, H represented harmony, M represented melody, and C represented complexity.

Leo tested his equation by applying it to his own compositions. He meticulously crafted pieces that balanced rhythm, harmony, and melody, ensuring that each element complemented the others. He tracked the reception of his music over several months and found that the equation accurately predicted the audience's enjoyment.

Encouraged by his success, Leo shared his findings with fellow musicians and composers. They began to use the Music Harmony Equation to guide their work, making adjustments to create more harmonious and captivating pieces. The city of Melodia thrived, with musicians producing compositions that resonated deeply with listeners.

Leo's story proved that math could be a powerful tool in music theory. His innovative approach inspired others to embrace mathematical principles in their musical pursuits. Melodia became a beacon of musical excellence, demonstrating that a balanced life, guided by math, could lead to lasting artistic achievement.

Compacting the Story

200 Words

In Melodia, mathematician Leo was fascinated by the relationship between math and music theory. He believed mathematical principles could unlock new dimensions in music composition. Leo analyzed classical compositions, identifying patterns in rhythm, harmony, and melody.

He created the Music Harmony Equation: $M=(R+H+M)C$, where M represented the musical score, R represented rhythm, H represented harmony, M represented melody, and C represented complexity.

Leo tested his equation by applying it to his own compositions. He crafted pieces that balanced rhythm, harmony, and melody. He tracked the reception of his music and found the equation accurately predicted the audience's enjoyment.

Encouraged by his success, Leo shared his findings with fellow musicians. They used the Music Harmony Equation to guide their work, creating more harmonious pieces. Melodia thrived, with musicians producing compositions that resonated deeply with listeners.

Leo's story proved math could be a powerful tool in music theory. His approach inspired others to embrace mathematical principles in their musical pursuits. Melodia became a beacon of musical excellence.

150 Words

In Melodia, mathematician Leo was fascinated by the relationship between math and music theory. He believed mathematical principles could unlock new dimensions in music composition. Leo analyzed classical compositions, identifying patterns in rhythm, harmony, and melody.

He created the Music Harmony Equation: $M=(R+H+M)C$, where M represented the musical score, R represented rhythm, H represented harmony, M represented melody, and C represented complexity.

Leo tested his equation by applying it to his own compositions. He crafted pieces that balanced rhythm, harmony, and melody. He tracked the reception of his music and found the equation accurately predicted the audience's enjoyment.

Encouraged by his success, Leo shared his findings with fellow musicians. They used the Music Harmony Equation to guide their work, creating more harmonious pieces. Melodia thrived, with musicians producing compositions that resonated deeply with listeners.

100 Words

In Melodia, mathematician Leo was fascinated by the relationship between math and music theory. He believed mathematical principles could unlock new dimensions in music composition. Leo analyzed classical compositions, identifying patterns in rhythm, harmony, and melody.

He created the Music Harmony Equation: $M=(R+H+M)C$.

Leo tested his equation by applying it to his own compositions. He crafted pieces that balanced rhythm, harmony, and melody. He tracked the reception of his music and found the equation accurately predicted the audience's enjoyment.

Encouraged by his success, Leo shared his findings. Melodia thrived, with musicians producing compositions that resonated deeply with listeners.

50 Words

In Melodia, mathematician Leo used math to analyze music. He created the Music Harmony Equation: $M=(R+H+M)C$. Leo tested it and shared his findings. Melodia thrived, with musicians producing compositions that resonated deeply with listeners.

25 Words

Leo created the Music Harmony Equation: $M=(R+H+M)C$. Melodia thrived, with musicians producing compositions that resonated deeply with listeners.

20 Words

Leo created the Music Harmony Equation. Melodia thrived, with musicians producing compositions that resonated deeply with listeners.

15 Words

Leo created the Music Harmony Equation. Melodia thrived.

10 Words

Leo's equation analyzed music. Melodia thrived.

7 Words

Leo's equation analyzed music, increased excellence.

5 Words

Equation analyzed music, excellence.

4 Words

Equation analyzed music.

3 Words

Analyzed music.

2 Words

Music analyzed.

1 Word

Analyzed.

0 Words

(Silence)

Explanation

The story illustrates how math can be used to understand and create music. By compacting the story, we distill its essence, showing how the core message remains powerful even in its simplest form.

Detailed Comments

- **Clarity**: The story and its compacted versions maintain a clear message about the power of math in music theory.

- **Dynamism**: The use of mathematical principles adds a dynamic element to the narrative.

- **Efficiency**: The compacted versions efficiently convey the core message without losing its essence.

- **Effectiveness**: Each version effectively communicates the relationship between math and music.

- **Realism**: The story is grounded in realistic applications of math in musical analysis.

Series of Opinions

1. Math can be a powerful tool for musical analysis.
2. Mathematical principles can simplify complex tasks.

3. Compacting stories can reveal their core essence.
4. Math can guide practical ways to achieve musical excellence.
5. Practical math can inspire creativity in music.

Sonnets

Sonnet 1: The Music Harmony Equation

In Melodia, where notes and numbers blend, A mathematician sought a way to mend, His love for music, formulas, and more, To find the harmony he could not ignore.

He crafted equations to light the way, To balance rhythm, melody, love each day. The Music Harmony Equation was his key, To finding joy and living happily.

His town embraced his method, clear and bright, And found their songs were filled with pure delight. With math and wisdom, they did find their way, To musical joy that lasts both night and day.

His story proves that numbers hold the key, To living life with notes and harmony.

Sonnet 2: Leo's Mathematical Quest

In Melodia, Leo's quest began, To find a way to help his fellow man. With math and data, he did seek to find, A method to bring peace to troubled mind.

He gathered experts, analysts, and more, To build a model, solid to the core. The Music Harmony Equation was their aim, A formula to bring melody's gentle flame.

With careful thought and rigorous design, They crafted tools to help the notes align. The town embraced his method, clear and bright, And found their songs were filled with pure ...

***

Highlights the relationship of math to a sailing adventure around the world:

Microfiction Story (299 words)

In the bustling port city of Nautica, a mathematician named Leo dreamed of sailing around the world. He believed that math could guide him on this epic adventure. Leo began by meticulously planning his route, using geometry to calculate the shortest paths between continents and trigonometry to navigate the open seas.

He created a mathematical model to optimize his journey, taking into account factors such as wind patterns, ocean currents, and weather conditions. Leo called it the Sailing Success Equation: $S=(W+C+T)R$, where S represented the sailing score, W represented wind patterns, C represented ocean currents, T represented travel time, and R represented risk.

Leo tested his equation by simulating various routes and scenarios. He adjusted his plans based on the results, ensuring that his journey would be both efficient and safe. With his calculations complete, Leo set sail on his adventure.

As he navigated the vast oceans, Leo relied on his mathematical skills to make real-time adjustments to his course. He used algebra to calculate fuel consumption, calculus to predict weather changes, and probability to assess risks. His journey was filled with breathtaking sights, from the icy fjords of Norway to the vibrant coral reefs of Australia.

After months at sea, Leo successfully completed his circumnavigation, returning to Nautica as a hero. His story proved that math could be a powerful tool in navigating the world. Leo's innovative approach inspired others to embrace mathematical principles in their own adventures. Nautica became a hub of exploration and discovery, demonstrating that a balanced life, guided by math, could lead to extraordinary achievements.

Compacting the Story

200 Words

In Nautica, mathematician Leo dreamed of sailing around the world. He believed math could guide him on this epic adventure. Leo meticulously planned his route, using geometry to calculate the shortest paths and trigonometry to navigate the seas.

He created the Sailing Success Equation: $S=(W+C+T)R$, where S represented the sailing score, W represented wind patterns, C represented ocean currents, T represented travel time, and R represented risk.

Leo tested his equation by simulating various routes and scenarios. He adjusted his plans based on the results, ensuring his journey would be efficient and safe. With his calculations complete, Leo set sail.

As he navigated the oceans, Leo relied on his mathematical skills to make real-time adjustments. He used algebra to calculate fuel consumption, calculus to predict weather changes, and probability to assess risks. His journey was filled with breathtaking sights.

After months at sea, Leo completed his circumnavigation, returning to Nautica as a hero. His story proved math could be a powerful tool in navigating the world. Leo's approach inspired others to embrace math in their adventures.

150 Words

In Nautica, mathematician Leo dreamed of sailing around the world. He believed math could guide him on this epic adventure. Leo meticulously planned his route, using geometry to calculate the shortest paths and trigonometry to navigate the seas.

He created the Sailing Success Equation: $S=(W+C+T)R$, where S represented the sailing score, W represented wind patterns, C represented ocean currents, T represented travel time, and R represented risk.

Leo tested his equation by simulating various routes and scenarios. He adjusted his plans based on the results, ensuring his journey would be efficient and safe. With his calculations complete, Leo set sail.

As he navigated the oceans, Leo relied on his mathematical skills to make real-time adjustments. His journey was filled with breathtaking sights.

After months at sea, Leo completed his circumnavigation, returning to Nautica as a hero. His story proved math could be a powerful tool in navigating the world.

100 Words

In Nautica, mathematician Leo dreamed of sailing around the world. He believed math could guide him. Leo meticulously planned his route, using geometry and trigonometry.

He created the Sailing Success Equation: \(S = \frac{(W + C + T …

CHAPTER 29

Example: "Last year, I had an experience that completely changed my outlook on life. I wanted to finish a marathon to prove to my kids that anyone can do it. One week before the big run, I slipped and injured my wrist. Completely shocked, I stood up, looking at my wrist which was now covered in blood. I thought, 'How can this happen to me just days before the marathon?' Despite the injury, I completed the marathon, proving to my kids that perseverance pays off."

Less than 299 Words

Focus on the essential elements: catchy opener, goals, obstacles, emotions, thoughts, and resolution.

Example: "I wanted to finish a marathon to prove to my kids that anyone can do it. One week before the big run, I slipped and injured my wrist. Despite the injury, I completed the marathon, proving to my kids that perseverance pays off."

Less than 150 Words

Condense the story further, keeping only the most critical elements.

Example: "I wanted to finish a marathon to prove to my kids that anyone can do it. One week before the run, I injured my wrist. Despite the injury, I completed the marathon, proving that perseverance pays off."

Less than 75 Words

Boil down to the core message.

Example: "I wanted to finish a marathon to prove to my kids that anyone can do it. Despite injuring my wrist a week before, I completed the marathon, showing that perseverance pays off."

Less than 25 Words

Capture the essence in a single sentence.

Example: "Injured a week before, I finished the marathon to prove to my kids that perseverance pays off."

Less than 19 Words

Further distill the message.

Example: "Injured before the marathon, I finished to show my kids perseverance pays off."

6 Words

Example: "Injured, finished marathon, proved perseverance."

5 Words

Example: "Injured, finished marathon, proved perseverance."

4 Words

Example: "Injured, finished, proved perseverance."

3 Words

Example: "Finished, proved perseverance."

2 Words

Example: "Proved perseverance."

1 Word

Example: "Perseverance."

0 Words

Example: Silence can sometimes be the most powerful statement.

Math Functions Representation

1. **Clarity**: ($f(x) = \text{clear}(x)$)
2. **Dynamism**: ($g(x) = \text{dynamic}(x)$)

3. **Efficiency**: (h(x) = \text{efficient}(x))

4. **Effectiveness**: (i(x) = \text{effective}(x))

5. **Realism**: (j(x) = \text{realistic}(x))

Combining these functions: [S(x) = f(x) + g(x) + h(x) + i(x) + j(x)]

This function (S(x)) represents an exceptional story that embodies clarity, dynamism, efficiency, effectiveness, and realism.

CHAPTER 30

And on your next truly great adventure, here are a few easy folk guitar songs to play, along with their chords that tell an excellent math function story:

1. **Brown Eyed Girl** by Van Morrison - G, C, D
2. **Down in the Valley** by Johnny Cash - C, G
3. **Stand By Me** by Ben E. King - G, Em, C, D
4. **Take Me Home, Country Roads** by John Denver - G, D, Em, C
5. **Hallelujah** by Jeff Buckley - C, Am, F, G, Em
6. **Blowin' in the Wind** by Bob Dylan - G, C, D
7. **The Times They Are A-Changin'** by Bob Dylan - G, C, D
8. **If I Had a Hammer** by Pete Seeger - G, C, D
9. **This Land Is Your Land** by Woody Guthrie - G, C, D
10. **The House of the Rising Sun** by The Animals - Am, C, D, F, E

11. **Man of Constant Sorrow** by The Soggy Bottom Boys - G, C, D

12. **Don't Think Twice, It's All Right** by Bob Dylan - C, G, Am, F

13. **I Saw the Light** by Hank Williams - G, C, D

14. **Big Rock Candy Mountain** by Harry McClintock - C, G, F

15. **Get Rhythm** by Johnny Cash - G, C, D

16. **Just Breathe** by Pearl Jam - C, G, Am, F

17. **Sunshine on My Shoulders** by John Denver - G, C, D

18. **Keep on the Sunny Side** by The Carter Family - G, C, D

19. **Old Time Religion** by Traditional - G, C, D

20. **I Walk the Line** by Johnny Cash - G, C, D

21. **Nine Pound Hammer** by Merle Travis - G, C, D

22. **Will the Circle Be Unbroken** by The Carter Family - G, C, D

23. **There Is a Time** by The Dillards - G, C, D

24. **Salvation Song** by The Avett Brothers - G, C, D

25. **Ring of Fire** by Johnny Cash - G, C, D

26. **Folsom Prison Blues** by Johnny Cash - G, C, D

27. **Amazing Grace** by Traditional - G, C, D

28. **Hallelujah** by Leonard Cohen - C, Am, F, G, Em

29. **Wagon Wheel** by Old Crow Medicine Show - G, D, Em, C

30. **Leaving on a Jet Plane** by John Denver - G, C, D

31. **Mr. Tambourine Man** by Bob Dylan - G, C, D

32. **City of New Orleans** by Arlo Guthrie - G, C, D

33. **Scarborough Fair** by Simon & Garfunkel - Am, G, Em, D

34. **Where Have All the Flowers Gone** by Pete Seeger - G, C, D

35. **500 Miles** by The Proclaimers - G, C, D

36. **Puff the Magic Dragon** by Peter, Paul and Mary - G, C, D

37. **Michael Row the Boat Ashore** by Traditional - G, C, D

38. **Tom Dooley** by The Kingston Trio - G, C, D

39. **Kumbaya** by Traditional - G, C, D

40. **This Little Light of Mine** by Traditional - G, C, D

41. **Oh Susanna** by Stephen Foster - G, C, D

42. **She'll Be Coming 'Round the Mountain** by Traditional - G, C, D

43. **Home on the Range** by Traditional - G, C, D

44. **Red River Valley** by Traditional - G, C, D

45. **Shenandoah** by Traditional - G, C, D

46. **Down by the Riverside** by Traditional - G, C, D

47. **Swing Low, Sweet Chariot** by Traditional - G, C, D

48. **When the Saints Go Marching In** by Traditional - G, C, D

49. **He's Got the Whole World in His Hands** by Traditional - G, C, D

50. **Rock-a-My-Soul** by Traditional - G, C, D

51. **Go Tell It on the Mountain** by Traditional - G, C, D

52. **Joshua Fit the Battle of Jericho** by Traditional - G, C, D

53. **We Shall Overcome** by Traditional - G, C, D

54. **Down in the River to Pray** by Traditional - G, C, D

55. **The Water Is Wide** by Traditional - G, C, D

56. **Wayfaring Stranger** by Traditional - G, C, D

57. **Danny Boy** by Traditional - G, C, D

Now, let's represent these songs as a series of math functions:

1. **Clarity**: ($f(x) = \text{clear}(x)$)
2. **Dynamism**: ($g(x) = \text{dynamic}(x)$)
3. **Efficiency**: ($h(x) = \text{efficient}(x)$)
4. **Effectiveness**: ($i(x) = \text{effective}(x)$)
5. **Realism**: ($j(x) = \text{realistic}(x)$)

Combining these functions: $[S(x) = f(x) + g(x) + h(x) + i(x) + j(x)]$

This function ($S(x)$) represents an exceptional story that embodies clarity, dynamism, efficiency, effectiveness, and realism.

***

Or consider:

Music at The Mansion: Mandolin Orange:

Tuesday May 28th, Mandolin Orange played a special set at the North Carolina Executive Mansion in Raleigh to help usher in a partnership between Come Hear North Carolina and the Americana Music Association (AMA)

Music at The Mansion: Mandolin Orange (youtube.com)

Mark Knopfler Barcelona 2019

Mark Knopfler Barcelona 2019 (youtube.com)

Steve Martin and the Steep Canyon Rangers: NPR Music Tiny Desk Concert

[Steve Martin and the Steep Canyon Rangers: NPR Music Tiny Desk Concert (youtube.com)](youtube.com)

Yet elsewhere, in a mansion event, the family seems stunned, as the crowd has an enormous buzz of guests, in various clusters.

***

So much so, it gives the complete impression of a spectacle, a show of shows function, filled with bright lights, and live chamber music, that includes Joseph Haydn String Quartet, in C, opus seventy-six, number three, the "Kaiser:" then other masterpieces:

Beethoven, the "Für Elise" …

… "Sonata Waldstein op 53" …

… "Sonata Opus 49, no. 2 in G major" …

***

… Frédéric François Chopin, "Raindrop Preludes, Op. 28: No. 15 in D-Flat Major"

***

… Joseph Haydn, "String Quartet in C, opus seventy-six, number three, Kaiser" …

… Wolfgang Amadeus Mozart, baptized, Johannes Chrysostomus Wolfgangus Theophilus Mozart …

… "Klavierstück, in F Major, K. 33B" …

… "Piano Concerto No. 19 K. 459" …

… "Rondo alla turca" …

… "Piano Concerto No. 21 K. 467, 2nd mov." …

… "Sonata No. 10 in C Major, K 330" …

… "Piano Sonata in G major, K. 283, 1st mov. Allegro" …

… "Sonata KV 545 DVT 2" …

… "Sonata KV 279, Mov 1" …

… "Rondo K 485" …

… "Sonata No 16 C major K 545 Barenboim" …